MOUNTAINS

The origins of the Earth's mountain systems

Also by Graham Park and published by Dunedin Academic Press

Introducing Geology – A Guide to the World of Rocks
(Second edition 2010)

Introducing Tectonics, Rock Structures and Mountain Belts
(2012)

The Making of Europe: A geological history
(2014)

Introducing Natural Resources
(2015)

For details of these and other Dunedin Earth and Environmental Sciences titles see
www.dunedinacademicpress.co.uk

MOUNTAINS

The origins of the Earth's mountain systems

Graham Park

EDINBURGH ◆ LONDON

Published by Dunedin Academic Press Ltd
Head Office: Hudson House, 8 Albany Street, Edinburgh EH11 3QB
London Office: The Towers, 352 Cromwell Tower, Barbican, EC2Y 8NB

www.dunedinacademicpress.co.uk

ISBNs
9781780460666 (Hardback)
9781780465791 (ePub)
9781780465807 (Kindle)

British Library Cataloguing in Publication data
A catalogue record for this book is available from the British Library

Typeset by Makar Publishing Production
Printed in Poland by Hussar Books

Contents

Sourced illustrations

The following illustrations are reproduced by permission:
Shutterstock: figures 2.1, 4.2, 4.8, 4.11, 4.14, 5.1, 5.6, 5.7, 5.9, 5.11, 6.3, 6.4, 6.8, 6.11, 6.13, 7.1, 8.1, 9.1, 9.4, 10.1, 10.8, 10.10, 11.1, 11.3, 11.5, 11.9, 11.10, 12.2, 12.5, 12.6, 12.9, 13.1, 13.7.
Science Photo Library: 7.4, 13.17.
Figure 2.2: Umbgrove, J.H.F. 1950. *Symphony of the Earth*, Martinus Nijhoff, The Hague.

The following illustrations have been adapted from published sources:
Figure 2.3: Du Toit, A.L. (1957) *Our Wandering Continents. An Hypothesis of Continental Drifting*. London: Oliver & Boyd.
Figure 2.4: Wegener, A. (1922) Die Enstehung der Kontinente und Ozeane. *Braunschweig: Friedrich Vieweg & Sohn.*
Figure 2.5: Du Toit, A.L. (1957) *Our Wandering Continents. An Hypothesis of Continental Drifting*. London: Oliver & Boyd.
Figure 2.6: Holmes, A. (1929) Radioactivity and earth movements. *Transactions of the Geological Society of Glasgow*, **18**, 559–606.
Figure 2.7: McElhinny, N.W. (1973) *Palaeomagnetism and plate tectonics*. Cambridge: Cambridge University Press.
Figure 2.8: Wyllie, P.J. (1976) *The way the Earth works*. New York: Wiley.
Figure 2.11: Larson, R.L. and Pitman, W.C. (1972) *Bulletin of the Geological Society of America*, **83**, 3645–3661.
Figure 3.1: Chadwick, P. (1962) Mountain-building hypotheses. In: S.K. Runcorn (ed.) *Continental drift*. New York: Academic Press.
Figure 3.3: Larson, R.L. and Pitman, W.C. (1972) *Bulletin of the Geological Society of America*, **83**, 3645–3661.
Figure 3.4: Uyeda, S. (1978) *The new view of the Earth*, San Francisco: Freeman.
Figure 3.5: Isacks, B., Oliver, J. and Sykes, L.R. (1968) Seismology and the new global tectonics. *Journal of Geophysical Research*, **73**, 18, 5855–5899.
Figure 3.6: Dewey, J. (1972) Plate tectonics. In: *Continents adrift and continents aground: readings from Scientific American*. San Francisco: Freeman.
Figure 3.7: Vine, F.J. and Hess, H.H. (1970) In: A.E. Maxwell (ed.) *The Sea*, v.4, New York: Wiley.
Figure 3.8C: Reston, T.J. (2007) The formation of non-volcanic rifted margins by the progressive extension of the lithosphere: the example of the West Iberian margin. From: G.D. Karner, G. Manatschcal and L.M. Pinheiro (eds) *Imaging, mapping and modelling continental lithosphere extension and breakup*. Geological Society, London, Special Publications, **282**, 77–110.
Figure 3.9: weebly.com/somali plate, via Wikimedia Commons.
Figure 3.10, 3.11: Dewey, J.F. and Bird, J. (1970) Mountain belts and the new global tectonics. *Journal of Geophysical Research*, **75**, 2625–2647.
Figure 3.13. Searle, M.P., Elliott, J.R. *et al.* (2011) Crustal-lithospheric structure and continental extrusion of Tibet. *Journal of the Geological Society, London*, **168**, 633–672.
Figure 4.3: Sans de Galdeano, C. (2000) Evolution of Iberia during the Cenozoic with special emphasis on the formation of the Betic Cordillera and its relation with the western Mediterranean. *Ciências da Terra (UNL), Lisboa* **14**, 9–24.
Figure 4.4: Ziegler, P. (1990) *Geological atlas of Western and Central Europe*. Shell Internationale Petroleum Maatschapij BV, 239pp.
Figure 4.5, 4.6: Handy, M.R., Schmid, S.M., Bousquet, R., Kissling, E. and Bernoulli, D. (2010) Reconciling plate-tectonic reconstructions of Alpine Tethys with the geological-geophysical record of spreading and subduction in the Alps. *Earth Science Reviews*, **102**, 121–158.
Figure 4.7, 4.9: (1) Puigdefàbregas, C., Muñoz, J.A. and Vergés, J. (1992) Thrusting and foreland basin evolution in the Southern Pyrenees. In: K. McClay (ed.) *Thrust tectonics*. . London: Chapman & Hall, 247–254. (2) Vergés, J. (1993) *Estudi geològic del vessant Sud del Pirineu Oriental i Central: Evolució en 3D*. Ph.D. thesis, University of Barcelona.
Figure 4.10: (1) Alonso-Chaves, F.M., Soto, J.I., Orozco, M., Kilias, A.A. and Tranos, M.D. (2004) Tectonic evolution of the Betic Cordillera: an overview. *Bulletin of the Geological Society of Greece*, **36**, 1598–1607. (2) Castro, J.M., Garcia, A., Gómez, J.J., Goy, A., Molina, J.M., Ruiz Ortiz, P.A. and Sopeña, A. (2009) Mesozoic successions of the Betic and Iberian ranges. In: A. Garcia-Cortés, J.A. Villar, J.P. Suarez-Valgrande and C.I.S. González, *Spanish geological frameworks and geosites*, Instituto Geológico y Minerò de España, Madrid, 73–90.
Figure 4.12: (A): (1) Azañón, J.M., Galindo-Zalvidar, J., Garcia-Dueñas, V. and Jabaloy, A. (2002) Alpine tectonics II: Betic Cordillera and Balearic Islands. In: W. Gibbons and M.T. Moreno, *The geology of Spain*. The Geological Society, London. (2) Morales, J., Serrano, I., Jabaloy, A. et. al. (1999) Active continental subduction beneath the Betic Cordillera and the Alboran Sea. *Geology*, **27**, 735–538. (B) Banks, C.J. and Warburton, J. (1991) Mid-crustal detachment in the Betic system of Southeast Spain. *Tectonophysics*, **191**, 275–289.
Figure 5.2: Argand, E. (1916) Sur l'arc des Alpes occidentales. *Eclogae Geologicae Helvetiae*, **14**, 145–191.
Figure 5.3: Zeck, H.P. (1999) Alpine plate kinematics in the western Mediterranean: a westwards-directed subduction regime followed by slab roll-back and slab detachment. In: B. Durand, L. Jolivet, F. Horvath and M. Séranne (eds) *The Mediterranean basins: Tertiary extension within the Alpine Orogen*. Geological Society of London, Special Publications, **156**, 109–120.
Figure 5.4: (A, C): Schmid, S.M., Fügenschuh, B., Kissling, E. and Schuster, R. (2004) Tectonic map and overall architecture of the Alpine orogeny. *Eclogae geologicae Helvetica*, **97**, 93–117; (B): Handy, M.R., Schmid, S.M., Bousquet, R., Kissling,

E. and Bernoulli, D. (2010) Reconciling plate-tectonic reconstructions of Alpine Tethys with the geological–geophysical record of spreading and subduction in the Alps. *Earth Science Reviews*, **102**, 121–158.

Figure 5.5: (A) Pfiffner, A. (2014) *Geology of the Alps*, Chichester: Wiley Blackwell. (B) Dietrich, D. and Song, H. (1984) Calcite fabrics in a natural shear environment, the Helvetic nappes of Switzerland. *Journal of Structural Geology*, **6**, 19–32.

Figure 5.8: Pfiffner, A. (2014) *Geology of the Alps*, Chichester: Wiley Blackwell.

Figure 5.10: Patacca, E. and Scandone, P. (2007) Geology of the Southern Apennines. *Bollettino del Societa Geologia Italiana*, Special Issue 7, 75–119.

Figure 5.12: Carminati, E. and Doglioni, C. (2012) Alps vs. Apennines: the paradigm of a tectonically asymmetric Earth. *Earth-Science Reviews*, **112**, 67–96.

Figures 6.1: Okay, A.I. (2000) Geology of Turkey: a synopsis, *Anschnitt*, **21**, 19–42.

Figure 6.2: Márton, E., Tischler, M., Csontos, L., Fügenschuh, B. and Schmid, S.M. (2007) The contact zone between the ALCAPA and Tisza–Dacia mega-tectonic units of Northern Romania in the light of new palaeomagnetic data. *Swiss Journal of Geosciences*, **100**, 1–16.

Figure 6.5, 6.6: Tari, V. (2002) Evolution of the northern and western Dinarides: a tectonostratigraphic approach. *EGU Stephan Mueller Special Publication* Series 1, 223–236.

Figure 6.7: Degnan, P.J. and Robertson, A.H.F. (2006) Synthesis of the tectonic–sedimentary evolution of the Mesozoic–Early Cenozoic Pindos Ocean: evidence from the NW Peloponnese, Greece. In: A.H.F. Robertson and D. Mountrakis (eds) (2006) *Tectonic development of the Eastern Mediterranean Region*. Geological Society, London, Special Publications, **260**, 467–491.

Figure 6.9: Robertson, A.F., Parlak, O, and Ustaömer, T. (2009) Mélange genesis and ophiolite emplacement related to subduction of the northern margin of the Tauride–Anatolide continent, central and western Turkey. In: D.J.J. Van Hinsbergen, M.A. Edwards and R. Govers (eds) *Collision and collapse at the Africa–Arabia–Eurasia subduction zone*. The Geological Society, London, Special Publications, **311**, 9–66.

Figure 6.10: (A) Okay, A.I. (2000) Geology of Turkey: a synopsis, *Anschnitt*, **21**, 19–42. (B) Dilek, Y. and Altunkaynak, S. (2009) Geochemical and temporal evolution of Cenozoic magmatism in western Turkey: mantle response to collision, slab break-off, and lithosphere tearing in an orogenic belt. In: D.J. Van Hinsbergen, M.A. Edwards and R. Glover (eds) *Collision and collapse at the Africa–Arabia–Eurasia subduction zone*. Geological Society, London, Special Publications, **311**, 213–233.

Figure 6.12: Okay, A.I. (2000) Geology of Turkey: a synopsis, *Anschnitt*, **21**, 19–42.

Figure 6.15: Adamia, S., Zakariadze, G., Chkhotua, T., Sadradze, N., Tsereteli, N., Chabukiani, A. and Gventsadze, A. (2011) Geology of the Caucasus: a review. *Turkish Journal of Earth Sciences*, **20**, 489–544.

Figure 7.2: Paul, A., Hatzfeld, D., Kaviani, A. and Péquegnat, C. (2010) Seismic imaging of the lithospheric structure of the Zagros mountain belt, (Iran). In: P. Leturmy and C. Robin (eds) *Tectonic and stratigraphic evolution of Zagros and Makran during the Mesozoic–Cenozoic*. Geological Society, London, Special Publications, **330**, 5–18.

Figure 7.3: Regard, V., Hatzfield, D., Molinaro, M., Aubourg, C., Bayer, R., Bellier, O., Yamini-Fard, F., Peyret, M. and Abassi, M. (2010) The transition between Makran subduction and the Zagros collision: recent advances in its structure and active deformation. In: P. Leturmy and C. Robin (eds) *Tectonic and stratigraphic evolution of Zagros and Makran during the Mesozoic–Cenozoic*. Geological Society, London, Special Publications, **330**, 43–64.

Figure 7.5, 7.6: McCall, G.J.H. and Kidd, R.G.W. (1982) The Makran, Southeastern Iran: the anatomy of a convergent plate margin active from Cretaceous to Present. In: *Trench–forearc geology: sedimentation and tectonics on modern and ancient active plate margins*. Geological Society, London, Special Publications, **10**, 387–397.

Figure 7.7: Platt, J.P., Leggett, J.K., Young, J., Raza, H. and Alam, S. (1985) Large-scale sediment underplating in the Makran accretionary prism, southwest Pakistan. *Geology*, **13**, 507–511.

Figure 7.8: Mahmood, S.A. and Gloaguen, R. (2012) Appraisal of active tectonics in Hindu Kush: Insights from DEM derived geomorphic indices and drainage analysis. *Geoscience Frontiers*, **3** (4), 407–428.

Figure 8.2: (1) Molnar, P. and Tapponnier, P. (1975) Cenozoic tectonics of Asia: effects of a continental collision. *Science*, **189** (4201), 419–426. (2) Searle, M.P., Law, R.D. and Jessup, M.J. (2006) Crustal structure, restoration and evolution of the Greater Himalaya in Nepal–South Tibet: implications for channel flow and ductile extrusion of the middle crust. In: R.D. Law, M.P. Searle and L. Godin (eds) *Channel flow, ductile extrusion and exhumation in continental collision zones*. Geological Society, London, Special Publications, **268**, 355–378.

Figure 8.3: NASA image.

Figure 8.4: Molnar, P. and Tapponnier, P. (1975) Cenozoic tectonics of Asia: effects of a continental collision. *Science*, **189** (4201), 419–426.

Figure 8.5: Harrison, T.M. (2006) Did the Himalayan crystallines extrude partially molten from beneath the Tibetan Plateau? In: R.D. Law, M.P. Searle and L. Godin (eds) *Channel flow, ductile extrusion and exhumation in continental collision zones*. Geological Society, London, Special Publications, **268**, 237–254.

Figure 8.6: Butler, R.W.H. and Prior, D.J. (1988) Tectonic controls on the uplift of the Nanga Parbat Massif, Pakistan Himalayas. *Nature*, **333**, 247–250.

Figure 8.8, 8.9: Searle, M.P., Elliott, J.R., Phillips, R.J. *et al.* (2011) Crustal–lithospheric structure and continental extrusion of Tibet. *Journal of the Geological Society, London*, **168**, 633–672.

Figure 8.10: Oil and Natural Gas Corporation, India, via Wikimedia Commons.

Figure 9.2: Metcalfe, I. (2011) Palaeozoic–Mesozoic history of SE Asia. In: R. Hall, M.A. Cottam and M.E.J. Wilson (eds) *The SE Asian gateway: history and tectonics of the Australia–Asia collision*. Geological Society, London, Special Publications, **355**, 7–35.

Figure 9.3: Hall, R. (2011) Australian–SE Asia collision: plate tectonics and crustal flow. In: R. Hall, M.A. Cottam and M.E.J. Wilson (eds) *The SE Asian gateway: history and tectonics of the Australia–Asia collision.* Geological Society, London, Special Publications, **355**, 75–109.

Figure 9.5: Kopp, H. (2011) The Javas convergent margin. In: R. Hall, M.A. Cottam and M.E.J. Wilson, M.E.J. (eds) *The SE Asian gateway: history and tectonics of the Australia–Asia collision.* Geological Society, London, Special Publications, **355**, 111–137.

Figure 9.6: Granath, J.W., Christ, J.M., Emmet, P.A. and Dinkelman, M.G. (2011) Pre-Cenozoic sedimentary section and structure as reflected in the JavaSPANTM crustal-scale PSDM seismic survey, and its implications regarding the basement terranes in the East Java Sea. In: R. Hall, M.A. Cottam and M.E.J. Wilson (eds) *The SE Asian gateway: history and tectonics of the Australia–Asia collision.* Geological Society, London, Special Publications, **355**, 53–74.

Figure 9.7: Hall, R. (2011) Australian–SE Asia collision: plate tectonics and crustal flow. In: R. Hall, M.A. Cottam and M.E.J. Wilson (eds) *The SE Asian gateway: history and tectonics of the Australia–Asia collision.* Geological Society, London, Special Publications, **355**, 75–109.

Figures 9.8, 9.9: Darman, H. (2014) *The Geology of Indonesia/ Banda Arc.* Wikibooks.

Figures 9.10, 9.11: Rangin, C. and Silver, E.A., *et al.* (1991) Neogene tectonics and evolution of the Celebes and Sulu basins: new insights from Leg 124 drilling. *Proceedings of the Ocean Drilling Program, Scientific Results,* **124**, 51–62.

Figure 10.2: Leat, P.T. and Larter, R.D. (2003) Intra-oceanic subduction systems: introduction. In: R.D. Larter and P.T. Leat (eds) *Intra-oceanic subduction systems: tectonic and magmatic processes.* Geological Society, London, Special Publications **219**, 1–17.

Figures 10.3, 10.4: Coates, G. (2002) *The rise and fall of the Southern Alps.* Christchurch, New Zealand: Canterbury University Press.

Figure 10.5: Segev, A., Ryabakov, M. and Mortimer, N. (2012) A crustal model for Zealandia and Fiji. *Geophysical Journal International,* **189**, 1277–1292.

Figure 10.6: Alataristarion (2006) via Wikimedia Commons.

Figure 10.7: Morrison, Sean (2014) *Geologic evolution of the Philippines.* https://geomorrison.files.wordpress.com.

Figure 10.9: Lallemand, S., Dominguez, S., Deschamps, A. and Liu, C-S. (2002) Arc–continent collision in Taiwan: new marine observations and tectonic evolution. *Geological Society of America,* Special Paper **358**, 189–213; Central Geological Survey of Taiwan, MOEA.

Figure 10.11: Taira, A., Ohara, S.R., Wallis, A., Ishiwatari, A. and Iryu, Y. (2016) Geological evolution of Japan: an overview. In: T. Moreno, S. Wallis, T. Kojima and W. Gibbons *The geology of Japan.* Geological Society, London.

Figure 10.12: Kojima, S. and 9 co-authors. (2016) Pre-Cretaceous accretionary complexes. In: T. Moreno, S. Wallis, T. Kojima and W. Gibbons *The geology of Japan.* Geological Society, London.

Figures 11.2: Moores, E.M. and Twiss, R.J. (1995) *Tectonics.* New York: Freeman.

Figure 11.4: Anon. (2011) Geological Surveys of the Yukon and British Columbia.

Figure 11.7: (1) Coney, P.J., Jones, D.L., and Monger, J.W.H. (1980) Cordilleran suspect terranes: *Nature,* **288**, 329–333. (2) Fitz-Diaz, E., Hudleston, P. and Tolson, G. (2011) Comparison of tectonic styles in the Mexican and Canadian Rocky Mountain Fold-thrust Belt. In: J. Poblet and R.J. Lisle (eds) *Kinematic evolution and structural styles of fold-thrust belts.* Geological Society, London, Special Publications, **349**, 149–167.

Figure 11.8: Moores, E.M. and Twiss, R.J. (1995) *Tectonics.* New York: Freeman.

Figures 12.1, 12.3: James, K.H. (2013) Caribbean geology: extended and subsided continental crust sharing history with eastern North America, the Gulf of Mexico, the Yucatán Basin and northern South America. *Geoscience Canada,* **40**, 1.

Figure 12.4: Westbrook, G.K. (1982) The Barbados ridge complex: tectonics of a mature fore-arc system. In: J.K. Leggett (ed.) *Trench–forearc geology: sedimentation and tectonics on modern and ancient active plate margins.* Geological Society, London, Special Publications, **10**, 357–372.

Figures 12.7, 12.8: (1) Moreno, T, and Gibbons, W. (2007) *The Geology of Chile.* The Geological Society, London. (2) Moores, E.M. and Twiss, R.J. (1995) *Tectonics.* New York: Freeman. Section 12.3: The Andes.

Figure 12.10: joannenova.com.au., via Wikimedia Commons.

Figure 12.11: Bulkeley, R. (2008) Aspects of the Soviet IGY. *Russian Journal of Earth Sciences,* **10**, ES1003.

Figure 13.3, 13.4: Searle, R. (2015) *Mid-ocean Ridges.* Cambridge: Cambridge University Press.

Figure 13.5: Uyeda, S. (1978) *The new view of the Earth.* San Francisco: Freeman.

Figures 13.6: topographic ocean-floor map by National Oceanographic and Atmospheric Administration (USA).

Figure 13.8: (1) Saemundsson, K. (1974) Evolution of the axial rifting zone in northern Iceland. *Bulletin of the Geological Society of America* **85**, 495–504. (2) Foulger, G.R. and Anderson, D.L. (2005) A cool model for the Iceland hotspot. *Journal of Volcanology and Geothermal research* **141**, 1–22.

Figure 13.11: Topographic ocean-floor map by National Oceanographic and Atmospheric Administration (USA).

Figure 13.12: Murton, B.J. and Rona, P.A. (2015) Carlsberg Ridge and Mid-Atlantic Ridge: comparison of slow-spreading centre analogues. *Deep Sea Research II, Topical studies in Oceanography,* **32**, 71–84.

Figure 13.13: Topographic ocean-floor map by National Oceanographic and Atmospheric Administration (USA).

Figure 13.14: Searle, R. (2015) *Mid-ocean Ridges.* Cambridge: Cambridge University Press.

Figure 13.15, 16: Topographic ocean-floor maps by National Oceanographic and Atmospheric Administration (USA).

Figure 13.18: Wilson, J.T. *(1963) A possible origin of the Hawaiian Islands. Canadian Journal of Physics.* **41***(6),* 863–870.

Figures 14.1, 14.2: Cocks, L.R.M. and Torsvik, T.H. (2006) European geography in a global context from the Vendian to the end of the Palaeozoic. In: D.G. Gee and R. Stephenson (eds) *European lithosphere dynamics.* Geological Society of London, Memoirs, **32**, 83–95.

Figure 14.3: (A) Dalziell, I.W.D. (1997) Neoproterozoic–Palaeozoic geography and tectonics: review, hypothesis, environmental speculation. *Geological Society of America Bulletin*, **109**, 16–42. (B) Cocks, L.R.M. and Torsvik, T.H. (2006) European geography in a global context from the Vendian to the end of the Palaeozoic. In: D.G. Gee and R. Stephenson (eds) *European lithosphere dynamics*. Geological Society of London, Memoirs, **32**, 83–95.

Figure 14.4: Gee, D., Juhlin, C., Pascal, C. and Robinson, P. (2010) Collisional orogeny in the Scandinavian Caledonides. *Geologiska Föreningen i Stockholm Förhandlingar*, **132**, 29–44.

Figure 14.5: Roberts, D. (2003) The Scandinavian Caledonides: event chronology, palaeogeographic settings and likely modern analogues. *Tectonophysics*, **365**, 283–299.

Figure 14.6: (1) Leslie, G., Smith, M. and Soper, N.J. (2008) Laurentian margin evolution and the Caledonian Orogeny: a template for Scotland and East Greenland. In: A.K. Higgins, J.A. Gilotti and M. P. Smith (eds) *The Greenland Caledonides: evolution of the northwest margin of Laurentia*. Geological Society of America, Memoir **202**, 307–343. (2) Dewey, J.F. and Shackleton, R.M. (1984) A model for the evolution of the Grampian tract in the early Caledonides and Appalachians. *Nature*, London, **312**, 115–121.

Figure 14.8: (A) Elliott, D. and Johnston, M.R.W. (1980) Structural evolution in the northern part of the Moine thrust zone. *Transactions of the Royal Society of Edinburgh: Earth Sciences*, **71**, 69–96. (B) Treagus, J.E. (2000) *Solid geology of the Schiehallion district*. Memoir of the British Geological Survey, HMSO.

Figure 14.9: (1, Africa): Michard, A., De Lamotte, D.F., Saddiqi, O. and Chalouan, A. (2008) An outline of the Geology of Morocco. In: A. Michard *et al. Continental Evolution: the Geology of Morocco*. Lecture Notes in Earth Sciences, **116**, Springer-Verlag, Berlin; Heidelberg. (2, N. America): US Geological Survey: Appalachian zones in the United States, via Wikimedia Commons.

Figure 14.10: Hickman, R.G., Vargo, R.J. and Altany, R.M. (2009) Structural style of the Marathon thrust belt, West Texas. *Journal of Structural Geology*, **31**, 900–909.

Figure 14.11: (1) Ballèvre, M., Bosse, V., Ducassou, C. and Pitra, P. (2009) Palaeozoic history of the Armorican Massif: models for the tectonic evolution of the suture zones. *Comptes Rendus Geoscience*, **341**, 174–201. (2) Martinez-Catalan, J.R., Aller, J., Alonso, J.L. and Bastida, F. (2009) The Iberian Variscan orogen. In: A. Garcia-Cortés, J.A. Villar, J.P. Suarez-Valgrande and C.I.S. Gonzálaez. *Spanish geological frameworks and geosites*, Instituto Geológico y Minerò de España, Madrid, 13–30.

Figure 14.12: Windley, B.F., Alexeiev, D., Xiao, W., Kröner, A. and Badarch, G. (2007) Tectonic models for accretion of the Central Asian Orogenic Belt. *Journal of the Geological Society, London*, **164**, 31–47.

Figure 14.13: (A) Juhlin C., Friberg M., Echtler H., Hismatulin T., Rybalka A., Green A.G. and Ansorge J. (1998) Crustal structure of the Middle Urals: results from the ESRU experiments, *Tectonics*, **17**(5), 710–725. (B) Berzin, R., Oncken, O., Knapp, J.H., Pèrez-Estaún, A., Hismatulin, T., Yunusov, N. and Lipilin, A. (1996) Orogenic evolution of the Ural Mountains: results from an integrated seismic experiment. *Science*, **274**, 220–222.

All other illustrations are by the author.

Preface

Mountains have always been a source of wonder and are among the most spectacular features of the natural world. My *Oxford English Dictionary* defines a mountain as 'a large abrupt natural elevation of the ground' but that hardly does justice to the magnificence of the great mountain chains that are such a dominant feature of our planet and have been a magnet for mountaineers, geologists and tourists alike for centuries.

It is not obvious to the non-geologist why mountains exist, or why they are so high relative to the general ground level of the continents. As recently as 1935, the famous physicist Sir Harold Jeffreys believed, along with many geologists of that period, that the Earth was shrinking due to internal cooling, and that the horizontal forces produced by accommodating to the decreasing area of the Earth's outermost shell were responsible for producing the great mountain ranges. Despite Alfred Wegener having promoted the theory of continental drift back in 1914, the contracting Earth theory was still widely held until the mid-twentieth century. Moreover, it may not be apparent why the great mountain ranges occur where they do, as distinct linear features, rather than being randomly distributed across the continents. It was only in the 1960s that plate-tectonic theory at last gave us a plausible mechanism for the formation of mountain belts.

The most obvious mountain chains today, such as the Himalayas, the Alps and the Andes, are situated at currently active plate boundaries. Others are the product of a plate collision that happened far back in the geological past, and have no present relationship to a plate boundary. These are much lower, with a generally more gentle relief, worn down through millennia of erosion by rain or ice. Many mountains are formed entirely by volcanic activity and, although also found along plate boundaries, frequently occur singly or in small groups. The most impressive of the volcanic mountains are almost completely hidden, forming the great ocean ridges that rival the Himalayas in scale.

The purpose of this book is to take the reader on a geological tour through the world's great mountain systems, examining in each case the plate-tectonic processes thought to be responsible for their evolution. The book is not intended as a comprehensive description of the geology of the Earth's mountains, which would be impossible in a single volume, but, in providing a general overview of the main mountain systems and their plate-tectonic contexts, it is hoped to reveal their grandeur and complexity, and the ingenuity of the mechanisms that have been advanced to explain them.

Graham Park, 2017

Acknowledgements

I am indebted to Professor John Winchester for many helpful comments and suggestions that have resulted in significant improvements to this book, and to my wife, Sylvia, for her unfailing support and for checking the manuscript for general readability.

1

Introduction

Mountains are created by two complementary processes – uplift and erosion. While the role of erosion in moulding the landscape has been generally understood by scientists since the work of James Hutton in the late eighteenth century, the mechanism causing uplift has been a puzzle until comparatively recently, and it is necessary to explain how and why this elevation occurs if we are to understand the origin of mountains.

Most mountains occur within relatively well-defined, narrow belts or mountain chains separated by wide expanses of much lower-lying ground. Their distribution is not random, but is caused by the now well-understood geological processes of plate tectonics. Some mountains mark the site of a former plate collision – where one continental plate has ridden up over another, resulting in a zone of highly deformed and elevated rocks. Others are essentially volcanic in origin – the volcanoes occurring either singly or in linear zones.

Hollow mountains

Early investigators were puzzled by the fact that the great mountain ranges such as the Himalayas appeared to be less heavy than their volume would suggest; that is, they did not exert the amount of gravitational attraction that a body of their size should have done – they behaved as if they were hollow! This was then explained by the discovery that more of a mountain belt occurs below the surface than above; this phenomenon is explained by the principle of 'isostasy', which describes a state of general gravitational equilibrium in which the extra weight of a mountain should be balanced by a deficiency of the denser material beneath, thus implying that denser material at a lower level must have been displaced to accommodate the additional mass of the less-dense material. The usual analogy made is with an iceberg floating in water, where most of the ice is submerged. Without this insight, it is impossible to explain why mountains can exist.

Thus, to the geologist, a mountain is more than just the topographical shape of the mountain itself, which is largely due to the way in which it has been carved out by erosion, but is part of a structure that is much larger and extends both vertically downwards and, apart from single volcanoes, also horizontally along the belt or mountain chain of which it forms a part. Moreover, the geological composition and structure of the rocks composing the mountain are an essential clue to its origin.

The terms 'mountain' and 'orogenic belt'

It is important, at the outset, to explain the difference in meaning between the terms 'mountain belt' and 'orogenic belt'. The former is used in a geographical sense and describes a topographic feature. The terms 'orogenic belt' or 'orogen' are used by geologists to describe a suite of geological features in which the presence of mountains, either now or in the geological past, is an integral and essential part.

The age of a mountain

While geologists now have a good understanding of how mountains are formed, this is by no means the case for the average non-geologist. A popular misconception is that the age of a mountain is represented by the age of the rocks it contains, and it is important to emphasise that the age is given by the date at which the mountain was elevated and eroded, and that this bears no simple relationship to the age of its included rocks. For example, the grand mountains of Assynt in Northwest Scotland are often referred to as 'the oldest mountains in Scotland' by virtue of the fact that they are composed of relatively old rocks (mostly late Precambrian). However, the age of these mountains is in fact determined by the period during which they were elevated and consequently sculpted by erosion – which was in this case considerably later, and culminated in substantial modification during the last Ice Age.

Structure of the book

Modern ideas on the origin of mountains evolved gradually over several centuries. Chapter 2 traces the historical development of these ideas and the contributions of the geologists who have played an important part in it, summarising the various mechanisms that have been proposed historically for the origin of mountains until the1960s, when our ideas were completely transformed by the plate-tectonic theory.

The historical development of the plate-tectonic theory, with its implications for mountain formation, is summarised in chapter 3, followed by an outline of the structure and composition of a well-studied example of an existing mountain belt, the Himalayas, providing a template which can be applied to the various examples that follow in the succeeding chapters.

The Earth's mountain systems may be divided into three categories: 1) the currently active Alpine–Himalayan and circum-Pacific belts; 2) the ocean ridge network; and 3) the older mountain belts that are no longer active.

The geologically recent Alpine–Himalayan mountain belt, described in Chapters 4–8, follows a set of collisional boundaries between (largely) continental plates. The circum-Pacific system, described in Chapters 9–12, follows the complex set of subduction zones that border the Western Pacific Ocean, and the American Cordilleran system in the Eastern Pacific.

Chapter 13 is devoted to the great submerged mountain chains of the deep oceans: these include currently active spreading ridges as well as those following active faults, inactive fracture zones and volcanic island chains.

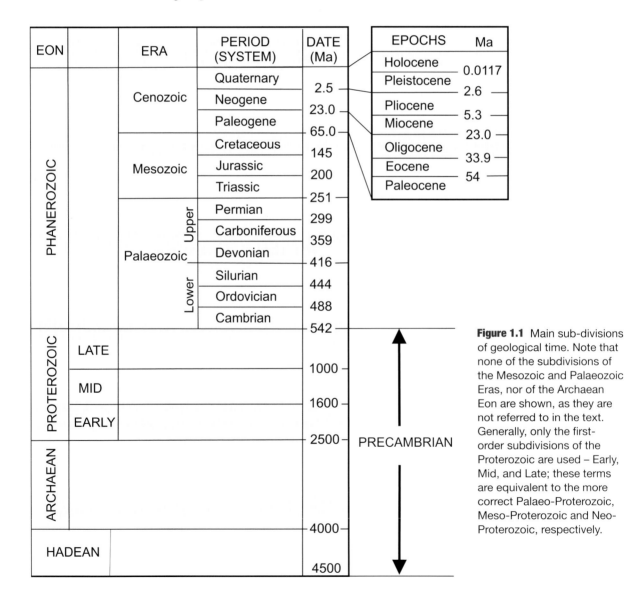

Figure 1.1 Main sub-divisions of geological time. Note that none of the subdivisions of the Mesozoic and Palaeozoic Eras, nor of the Archaean Eon are shown, as they are not referred to in the text. Generally, only the first-order subdivisions of the Proterozoic are used – Early, Mid, and Late; these terms are equivalent to the more correct Palaeo-Proterozoic, Meso-Proterozoic and Neo-Proterozoic, respectively.

The final chapter explores some of the older (pre-Mesozoic) mountain chains, including the Palaeozoic Caledonian and Hercynian orogenic belts.

Units of time and motion

The age of geological events is always given in Ma (million years) before present. Tectonic events are rarely known with much precision, and in any case usually span time periods of many millions of years. Most frequently they are quoted in terms of part of the geological timescale (Fig. 1.1) in terms such as 'mid-Cretaceous' or 'early Miocene', in which case the Table will give the exact time range referred to.

Rates of movement are usually in the range 1–20 millimetres per year and are quoted in the form: mm/a (mm per annum). These are invariably mean measurements or estimates based over several years, or in the case of plate movements, millennia.

2

Historical views

According to the ancient Greeks, the mountains were the home of the Gods – in Greek mythology, Mount Olympus, the highest mountain in Greece, was where Zeus held court with the other gods, and no human would dare to go there. There seems to have been little, if any, attempt by the ancient philosophers at understanding the true nature of mountains.

In both the Jewish and subsequently Christian traditions, mountains, like the other features of the Earth, were part of the divine creation, and for centuries it was futile (and at times dangerous) to question this. The serious scientific approach to the origin of mountains only began in the late eighteenth century with the work of James Hutton.

Hutton and Lyell
James Hutton (1726–1797)

Hutton was a Scottish scientist who had studied medicine in Edinburgh and Leiden, but abandoned a career as a physician to concentrate on his experiments in chemistry and on agricultural improvements on his farms. His detailed observations of soil being washed from the land and deposited into ditches and river beds, together with his knowledge of the rocks around Edinburgh, led him to the then revolutionary proposal that sedimentary rocks such as mudstone, sandstone and limestone were formed by observable processes of erosion of the land surface and the deposition of the derived material into a lake or sea. The sedimentary layers so produced were consolidated into hard rock by heat, and must then have been uplifted to their present positions on land where they can be observed now. This view contrasted with the prevailing theory, termed Neptunism, which held that all rocks (including crystalline igneous rocks such as granite!) were precipitated from the great biblical flood.

Hutton was familiar with the active volcanoes of Italy, and deduced that the rocks of Salisbury Crags in Edinburgh (Fig. 2.1) were of volcanic origin. He observed how veins of granite and dykes of basalt had penetrated into their host rocks, concluding that they

must have been molten, and younger than the host material. He proposed that the interior of the Earth was hot, and that this heat was responsible for the creation of new rock. As these processes were very gradual, he emphasised the great timespans necessary to explain the known geological record, and that in terms of the formation of the Earth, there was 'no vestige of a beginning, no prospect of an end'. This, again, was in conflict with the prevailing opinion, based on the contemporary religious orthodoxy, that the Earth was formed in a single event no more than a few thousand years ago.

Hutton's views were first published in the *Transactions of the Royal Society of Edinburgh* in 1788 but did not appear in book form until 1795. It was Hutton, known as the 'father of modern geology' who established geology as an independent science. However, although Hutton clearly understood that mountains were composed of a wide variety of rocks whose origins could be explained by sedimentary and igneous processes, he was unable to provide a satisfactory solution to the problem of how the mountains themselves originated.

Hutton had travelled to the Alps, and knew that layers of limestone containing the fossil shells of marine creatures, identical to those found in present-day seas, occurred high up in the Alps and other great mountain chains. He realised that some mechanism was required to elevate them from their original site of formation at the bottom of the sea to their present position. He was also aware of the widespread occurrence of igneous rocks such as granite in mountainous terrains, and knew that great heat was required to melt them. He therefore suggested that 'subterraneous fire' – a deep-seated heat source – caused the expansion necessary to elevate the mountains, arguing that if volcanoes such as Etna and Vesuvius could produce sizeable mountains purely as a result of internal heat, then surely the same process could also account for the much greater expansion necessary to explain the Alps. He noted that volcanoes occurred in many places around the margins of the Alps but not within the mountain chain itself, supposing that, if the

Figure 2.1 Salisbury Crags, Edinburgh. Hutton contended that these rocks were formed from a layer* of whinstone (basalt) of volcanic origin, i.e. derived from a magma, that had been injected into the surrounding sediments [*which we would now call a sill]. Shutterstock©christapper.

subterranean heat could not escape via molten lava from volcanoes, it may have provided a powerful mechanism for elevating the chain. He also believed that the intense 'contortions' (i.e. folding) of the strata observed in the Alps could be explained as part of the process of uplift of the massif.

Charles Lyell (1797–1875)

Hutton's ideas were popularised in two influential textbooks, *Principles of Geology* and *Elements of Geology* by Sir Charles Lyell (1797–1875), both of which ran to numerous editions spanning the period 1838 to 1865. Lyell, born of a wealthy Scottish family, and of independent means, was appointed Professor of Geology at University College, London and spent much of his time travelling in North America and Europe observing geological phenomena of all kinds. He was a close friend of Charles Darwin and influenced the latter's views on evolution. Lyell's work covered the whole field of geology

(the first edition of his '*Principles*' textbook extended to three volumes) but a major contribution was to make Hutton's views more accessible to the general scientific community. However, he made no further contribution to Hutton's ideas on the origin of mountains.

The contracting Earth theory

The idea that mountain chains were the result of the contraction of the outermost shell of the Earth has been attributed to the American scientist James Dwight Dana (1813–1895) but the theory of a contracting Earth was embraced by many other prominent geologists, including Eduard Suess (Suess, 1906) and Sir Archibald Geikie, whose textbook (Geikie, 1882) provides a good example of how the theory was used to explain the structure of mountains.

The basis of the contracting Earth theory was the belief that the Earth was slowly cooling from its originally hot molten state, and that the cooling of the molten interior gave rise to a shrinking of the solid outer shell or 'crust'– rather like the wrinkled skin of a dried-up apple! The future discovery of radioactive decay, and the consequent realisation that this additional heat source provided an ongoing supply of heat, meant that the cooling Earth theory had to be abandoned, but this was not generally acknowledged until the mid-twentieth century.

Archibald Geikie (1835–1924)

Sir Archibald Geikie was appointed as the first Professor of Geology at the University of Edinburgh in 1871 and subsequently as Director-General of the Geological Survey of Great Britain from 1881 to 1901; his *Textbook of Geology* (Geikie, 1882) was the standard guide for geology students for many decades. His views on mountain formation can be summed up in the following quotations from the third edition of his book, published in 1893.

> … the true mountain ranges of the globe … may be looked on as the crests of the great waves into which the crust of the Earth has been thrown.

And:

> These examples [i.e. the Alps, Rockies, etc. (author)] show that the elevation of mountains, like that of continents, has been occasional and sometimes paroxysmal. Long intervals elapsed, when a slow subsidence took place, but at last a period was reached when the descending crust, unable to withstand the accumulated lateral pressure, was forced to find relief by rising into mountain ridges.

And again:

> Geologists are now generally agreed that it is mainly to the effects of the secular contraction of our planet that the deformation and dislocation of the terrestrial crust are to be traced. The cool outer shell has sunk down upon the more rapidly contracting hot nucleus and the enormous lateral compression thereby produced has thrown the crust into undulations, and even into the most complicated corrugations.

Harold Jeffreys (1891–1989)

Sir Harold Jeffreys was a Fellow of St. John's College Cambridge, who published highly influential work on mathematics, geophysics and astronomy. His 1924 work, *The Earth, its origin, history and physical constitution* became a standard text on geophysics. Jeffreys was a prominent and effective opponent of the theory of continental drift because he believed that there was no known force strong enough to move the continents across the Earth's surface. His calculations of the strength of the Earth were based on the view that the Earth had cooled from an originally molten state, and that its strength was equivalent to that of the surface rocks. However, work by Sir Arthur Holmes and others in the 1920s on the heat generated by radioactive decay had shown that the interior of the Earth was much warmer than Jeffreys had thought – in fact, as we now know, close to the melting point of rocks at relatively shallow depths of about 50km in places. As rock becomes warmer, it also becomes significantly weaker, thus invalidating Jeffreys' calculations.

Jeffreys' views on the origin of mountains are well expressed in the following quotation from his later book, *Earthquakes and Mountains*, published in 1935.

> The elevation of a mountain system represents work done against gravity … At present the strength of the crust is preventing gravity from making the surface level, but when the mountains were formed it was aiding gravity in resisting the stresses that made them. To explain the origin of mountains we must provide sufficient stresses as will overcome both the strength of the Earth and gravity, for in a symmetrical body, both would act together in opposing any change of shape. The only agency that seems capable of supplying such stresses is contraction of the interior … a sinking crust has to acquire a shorter circumference to fit the new size of the interior.

The shortening needed was calculated by Jeffreys using the known crushing strength of granite, and indicated that a section of the crust could be compressed by 1/800th of its length before being crushed, and therefore it followed that the Earth's circumference of *c*.40,000km could be shortened by *c*.50km before failure (e.g. folding or faulting) occurs. From this he estimated that 70km shortening would be required to produce the Alps and 190km for the Himalayas. These figures contrast strikingly with those calculated from the actual structures within these mountain belts, as will be seen later.

From geosynclines to mountains

The concept of the geosyncline was first introduced by the American geologists James Hall and James Dwight

Dana in the mid-nineteenth century as a result of their studies of the Appalachian Mountains. The term was used to describe a large elongate basin within the crust that gradually deepened and became filled with sediment. The idea was later elaborated with the recognition of two distinct categories: the 'eugeosyncline', which included extensive vulcanicity and sediments typical of the continental slope and deep ocean, and the 'miogeosyncline', which contained sediments more typical of the continental shelf, and which lacked vulcanicity. The geosynclinal hypothesis of mountain formation supposed that, as the geosyncline filled up, it became unstable and collapsed under the combined influence of gravity and horizontal contraction. It subsequently became elevated to form a mountain chain. These ideas were widely accepted as an explanation for the origin of mountain belts until replaced by the plate-tectonic theory in the 1960s.

J.H.F. Umbgrove (1899–1954)
Umbgrove was a Dutch geologist whose ideas on the formation of mountain belts were strongly influenced by his knowledge of the Indonesian island chain, but he also had views on the origin of the Alps, which are illustrated by a lecture he gave to the *Société pour l'avancement des Sciences* at Geneva in 1948, and reproduced in a book, rather grandly titled *Symphony of the Earth* (Umbgrove, 1950).

Umbgrove describes what he imagines would be the result of a gradually deepened and filled geosyncline as follows.

> It is not difficult to imagine what must happen during a subsequent period of strong compression of the earth's crust. The contents of the furrow [i.e. the geosyncline] will become crumpled and folded. Some of the large folds will protrude over the border of the trough and slip onto the adjacent 'foreland' … In a more advanced stage the whole zone will tend to rise, for the thick pile of light sedimentary rocks will tend to re-establish equilibrium as soon as it gets the opportunity to do so, just as a submerged log pressed down in water will rise upwards as soon as the force which pressed it down is taken away. Consequently, from the elongated belt emerges a mountain range.

Umbgrove here, in his analogy of the floating log, is referring to what is known as the principle of isostasy. This holds that the Earth's crust is in a state of general gravitational equilibrium, such that a mass excess at the surface must be compensated by a mass deficiency beneath. Thus a mountain range must be supported beneath by a mass of material less dense than the surrounding lower-crustal rocks, such that the weight of that part of the crust would be equivalent to that of the adjacent borderlands in order to be gravitationally stable. The usual analogy is with an iceberg floating in water, where a large part of the ice is submerged. In the case of Umbgrove's mountain range, the excess of less dense material depressed into the denser substratum causes a gravitational imbalance that elevates the mountains, restoring them to a state of equilibrium.

The sequence of events envisaged by Umbgrove is illustrated in Figure 2.2. At an advanced stage of the depression of the trough (i.e. stage II in the figure), the

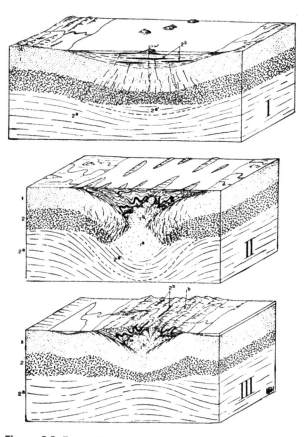

Figure 2.2 Formation of a mountain belt according to Umbgrove. Stage I: sediments accumulate in a gradually subsiding trough (geosyncline); II: under a combination of downward and lateral pressure, the base of the sediment pile collapses, allowing an influx of deep-crustal material (e.g. granite); III: gravitational equilibrium is restored with the elevation of the mountain range. Crustal layers: 1, granitic layer; 1a, granite batholith; 2, 'basic' layer; 2a, 'basic substratum'. © Umbgrove, 1950, with permission.

combination of gravitational and lateral pressure causes the disintegration of the lowest part of the trough, allowing the rise of granitic material from the lower crust. The final stage (III) is when gravitational equilibrium is restored by the uplift of the geosyncline to form the mountain belt.

Continental drift

Continental 'drift' was the name given to the concept of the relative movement of the continents around the Earth's surface, and was the first theory that offered the realistic possibility of explaining the localised contractions responsible for the creation of mountain belts such as the Alps. The contracting Earth theory advanced by geologists such as Suess and Jeffreys had eventually to be abandoned once it was generally accepted that the Earth was continually generating heat from radioactive sources, was not cooling down and therefore not shrinking. The idea of continental drift had been around even before 1912, when Alfred Wegener had proposed it. The American geologist F.B. Taylor had suggested in 1910 that 'crustal creep' of the continents could explain the shapes of the Alpine–Himalayan mountain chain – notably the convergence of India and Australia with Asia. However, the lack of an adequate mechanism for the movements, and of any convincing geological evidence in support of his ideas, meant that they received little recognition in comparison with Wegener's.

The reasons why continental drift took such a long time to be accepted as a valid hypothesis by the scientific community as a whole offer an interesting insight into the background and personalities of some of the key individuals in a debate that lasted for over forty years.

Alfred Wegener (1880–1930)

The theory of continental drift is usually attributed to Alfred Wegener, a distinguished German meteorologist who had undertaken four expeditions to Greenland to undertake meteorological investigations, during the last of which he tragically lost his life while undertaking a traverse of the icecap. He published his views on continental drift (in German) in 1912 in the scientific journal *Geologische Rundschau* and again in book form in 1922 (Wegener, 1922), to explain the numerous geometric and geological similarities between continents that are now separated by oceans. The continents of South America, Africa, India, Australia and Antarctica were shown to fit together in a supercontinent called Gondwanaland (originally named by Eduard Suess; now usually 'Gondwana'),

and North America and Eurasia into a second supercontinent called Laurasia (Fig. 2.3). These two supercontinents were joined in Central America to form a continuous worldwide landmass termed Pangaea (pronounced 'Pan-jee-a').

When this continental reconstruction was examined, many geological features shared by the separated continents could be explained. These include the presence of glacier-derived clays and glacial striations, which in their present positions cover about half the globe, but when restored to their presumed Gondwana fit, make a reasonably-sized polar ice cap. The distribution of other climatic indicators in rocks 200 million years old also makes sense when in the Gondwana fit; these include dune-bedded sandstones and salt deposits, which mark out two desert belts on either side of a central equatorial belt indicated by the presence of coal deposits and coral reefs suggestive of tropical conditions (Fig. 2.4). There are also similarities between many fossil land animals and

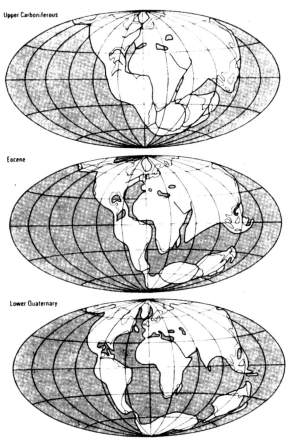

Figure 2.3 Wegener's reconstruction of the continents from the Carboniferous to the Quaternary. From Du Toit, 1937.

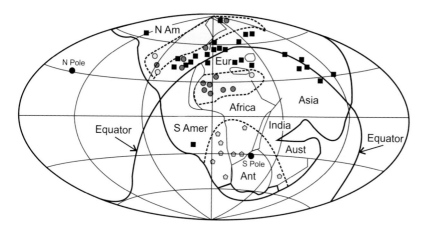

Figure 2.4 Gondwanaland in the Carboniferous. Climatic belts according to Wegener. Note: 1, the band of coal deposits parallel to the Carboniferous Equator (heavy black line); 2, desert areas (in yellow) marked by desert sandstones and evaporite deposits on either side of the equatorial belt; 3, large polar area showing evidence of glaciation. After Wegener, 1922.

○ ice ● evaporite (salt, gypsum) ○ desert sandstone ■ coal

plants that existed 200 million years ago in the different continents prior to the splitting up of the supercontinent, and this contrasts with the obvious differences that exist between the fauna and flora of the separate continents now. However, much of the detailed geological observation to support these ideas was undertaken later, by Alexander Du Toit.

Wegener believed that mountain belts such as the Alps and Himalayas could be explained by the effect of two continents converging during the amalgamation of Pangaea – Africa with Europe in the case of the Alps, and India with Asia, in respect of the Himalayas. The oceanic area partly enclosed by the super-continent was named the 'Tethys Ocean' by Eduard Suess (see above), who believed that the present-day Mediterranean Sea was a remnant of this former ocean, and that much of the former ocean-floor had been incorporated into the Alpine mountain belt during its formation. The circum-Pacific belt, which includes the American Cordilleran Chain and the Andes, was more difficult to explain by collision, since it required the Pacific Ocean crust to be strong enough to buckle the edge of the American continent, which would contradict Wegener's model of a weak ocean crust. Wegener had suggested that the continents were able to move from their original positions in Gondwana and Laurasia to their present ones because the oceanic crust was much weaker than the continents, which could somehow plough their way across it.

Wegener's lack of geological background and inability to provide sufficient 'geological' evidence to support his theory meant that his ideas received very little support, especially in the northern hemisphere, partly because they were initially published (in German) during the First World War. Wegener's book was not published in English until 1966. His views were communicated to a wider audience via a lecture he gave in New York in 1926, but were widely criticised. Many geologists, and particularly geophysicists, were influenced by Sir Harold Jeffreys' calculations of the strength of the Earth's crust, which we have already referred to, and opposed continental drift because they believed that the Earth's crust was too strong to allow the kind of behaviour that Wegener's theory required.

Wegener's ideas caused considerable debate among the geological community, but failed to obtain universal acceptance, mainly because of the lack of a convincing mechanism for the movements. It was not until the 1950s that advances in palaeo-magnetism produced enough evidence to convince geologists that the theory was correct.

Alexander du Toit (1878–1948)

Du Toit was a South African geologist who had studied mining in London and taught geology at Glasgow University in the early years of the twentieth century, but returned to South Africa and spent many years travelling extensively in southern Africa, South America and Australia gathering the detailed field evidence required to support the geological comparisons necessary to bolster the continental drift theory. In 1927, he published his studies comparing the geology of southern Africa with that of South America, and ten years later (Du Toit, 1937) his comprehensive work *Our Wandering Continents, An Hypothesis of Continental Drifting*, which gave much of the detailed evidence necessary to establish continental drift as a valid theory, to be taken seriously by a large proportion of the geological community.

In addition to providing many further examples of geological matches across the now severed edges of the southern continents, he was able to demonstrate much more convincing evidence for various climatic indicators in rocks of Carboniferous age in the southern Gondwana continents. He pointed out that the distribution of these climatic indicator rocks makes no sense in their present locations; for example, coals representing the product of equatorial forests now lie near the North Pole, and glacial deposits lie near the Equator!

It should be noted that the term 'continent' used in a geological sense includes, in addition to the landmass, areas of the adjacent sea bed – the continental shelf and continental slope – that are underlain by crust of continental, rather than oceanic, type. Du Toit showed that when the shape of the continents is adjusted to include these, a much better fit of the Gondwana continents is achieved (Fig. 2.5). His reconstruction of the pre-drift continents differed from Wegener's in that the northern grouping, Laurasia, was centred around the then North Pole and was separated from Gondwana by a wider Tethys Ocean.

Du Toit proved that much of the process of separation of the Gondwana continents must have taken place during the Mesozoic Era, whereas Wegener had thought that most of the movements had occurred in the Tertiary (Cenozoic) and continued into the Quaternary Period (i.e. the last 2.5Ma) as well. He also extended the drift mechanism to explain the older Paleozoic mountain belts such as the Caledonian belt, now severed by the Atlantic Ocean.

A warmer Earth – convection
Arthur Holmes (1890–1965)

Holmes pioneered the use of radioactivity in dating rocks, and published the first date using the uranium-lead method in 1911 shortly after graduating from Imperial College, London, where he continued to work on the subject, publishing *The Age of the Earth* in 1913, in which he estimated a date of 1600 million (1.6Ga) years for the oldest rocks then known. By the 1940s this date had been revised upwards to 4500+/-100Ma, approximately equivalent to the presently accepted date. In 1943 he was appointed Professor of Geology at Edinburgh University, where he remained until retirement. His *Principles of Physical Geology* (Holmes, 1944) was the standard textbook on the subject for many decades; the fourth edition, edited by P. McL. D. Duff, published in 1993, is still in use.

Holmes' work on radioactivity enabled him to demonstrate that the Earth must be much hotter than previous estimates, which had been based on the cooling Earth model. It must, therefore, also be much weaker than previously thought, and he suggested that the mantle could be capable of transferring heat from the interior to the surface by slow flow in the solid state. He visualised a system of convection currents (Fig. 2.6), the upper parts of which could carry continents across the Earth's surface. As well as providing the missing mechanism for continental drift, this idea was a major contributory factor in the development of the plate-tectonic theory, discussed in the following chapter.

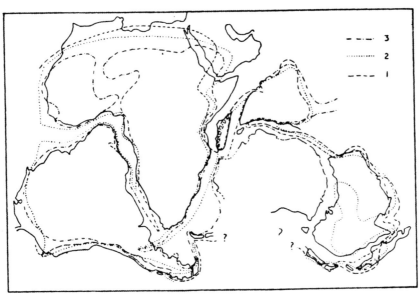

Figure 2.5 Gondwanaland according to Du Toit. Note that this reconstruction includes the continental shelves and is a much more accurate representation of the coastlines than Wegener's. The dotted and dashed lines represent the positions of the shorelines during (1) early Jurassic, (2) early Cretaceous and (3) late Cretaceous time. From Du Toit, 1937.

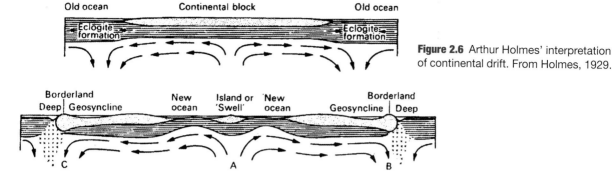

Figure 2.6 Arthur Holmes' interpretation of continental drift. From Holmes, 1929.

Palaeomagnetism

The debate on continental drift carried on until the 1950s, when work on palaeomagnetism (the study of the magnetism of old rocks) finally convinced most geoscientists that the continental drift theory was correct. The science of palaeomagnetism depends on the fact that certain rocks (e.g. basalts and iron-rich sandstones) acquire weak magnetisation from the contemporary geomagnetic field and behave as a kind of magnet. A sensitive magnetometer can measure this geomagnetic field to determine both the direction of magnetic north and the latitude of the magnetic field. This magnetic information is 'stored' within the rock, and unless it is disturbed by some subsequent event such as metamorphism, it will represent a 'fossil' magnetic field at the time when the rock was originally formed.

Work by a group of geophysicists at Cambridge University in the later 1950s obtained revolutionary results indicating that the apparent pole positions of rocks with ages varying from Jurassic to Quaternary showed widely different positions that formed a 'polar-wander curve' or track ending at the present pole. Moreover, the tracks for each continent were different, proving either that the Earth's magnetic field had behaved in a very strange and inexplicable way over the last 200 million years, or that the continents had indeed changed their relative positions with time. The Cambridge group included S.K. Runcorn, who subsequently headed the Geophysics Department at Newcastle University, and while there, summarised the results in a landmark paper published in a volume entitled *Continental Drift* (Runcorn, 1962).

Runcorn showed that, for example, the positions of magnetic north for 200 million-year-old rocks in different continents plotted in different places (Fig. 2.7A). However, when the continents were fitted together in their presumed original positions according to continental drift theory, the locations of the magnetic north poles coincided (Fig. 2.7B). This was convincing proof that the continental drift theory was in essence correct, and because the evidence was supported by respected geophysicists, managed to convert most of the doubters of the geological fraternity.

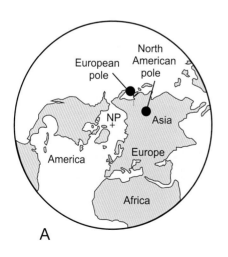

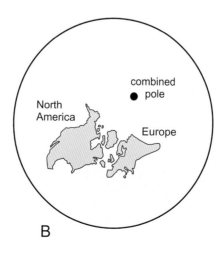

A   B

Figure 2.7 Palaeomagnetic evidence for continental drift. **A.** North Pole positions for Europe and Africa in 200Ma old rocks. **B.** Position of combined pole when continents are moved into their pre-drift positions. After McElhinny, 1973.

Sea-floor spreading
Topography of the ocean floor

The next stage in the evolution of ideas came from studies of the ocean floor, where mapping by various remote-sensing techniques had revealed a topography that was as varied as that of the continents. The generally even ocean floor is interrupted by a system of great ridges and deep, narrow trenches (Fig. 2.8). The ridges are typically between 1000 and 2000 kilometres wide and rise to as much as 3 kilometres from the ocean floor. They form a continuous network, one branch of which runs from the Arctic along the centre of the Atlantic Ocean (the Mid-Atlantic Ridge) to join a second branch that completely surrounds Antarctica and crosses the Pacific Ocean towards the coast of Mexico, sending two branches into the Indian Ocean. The trenches are much narrower (typically around 100 kilometres wide) but extend to depths of up to 11 kilometres below sea level. They form generally curved linear features on the ocean-ward side of island chains around the western Pacific Ocean, the eastern Indian Ocean, the Caribbean, and along the Pacific coast of America. Much of this information had been available even in Wegener's time, but its significance was not understood until the sea-floor spreading idea at last gave a believable explanation. Extensive work by American oceanographers based at Scripps Institute in California and Lamont-Docherty at Columbia University, New York, produced a greatly expanded knowledge of the topographic detail of the ridge–trench system, which will be described in detail in chapter 13.

H.H. Hess (1906–1969)

The breakthrough in understanding that finally solved the problem of a mechanism for plate movements came from H.H. Hess of Princeton University, who proposed that the ridges were underlain by hot, rising mantle currents and the deep-ocean trenches by cool, descending currents; this he called 'sea-floor spreading'. First published in a Report of the (US) Office of Naval Research in 1960, his theory was explained in detail in 1962 (Hess, 1962), and is also explained in the chapter by R.S. Dietz in Runcorn's book on Continental Drift (Runcorn, 1962). Much of the objection to Wegener's ideas on continental drift had centred on the failure to visualise how a continent could plough across a static ocean crust. However, this objection was countered by Hess's proposal that the ocean crust itself was mobile, and behaved like a giant conveyor belt, rising at the ridges and moving sideways towards the deep-ocean trenches, where it descended (Fig. 2.9). In other words, both continents and oceans were mobile rather than static.

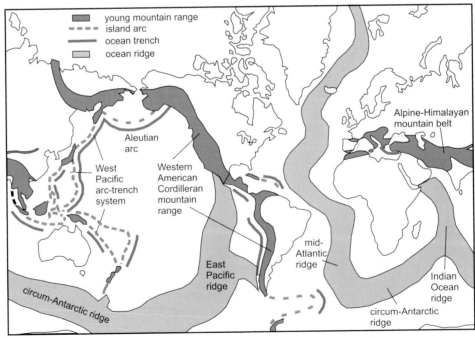

Figure 2.8 Principal topographic features of the Earth. Note the size of the ocean ridges compared with the continental mountain belts and the narrow ocean trenches. After Wyllie, 1976.

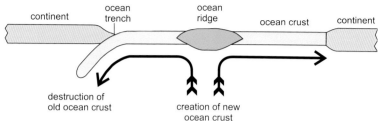

Figure 2.9 Sea-floor spreading. New ocean crust is created at ocean ridges and destroyed at ocean trenches by flow within the mantle.

Dating of the ocean floor

Wegener and Du Toit had thought that the ocean ridges marked the lines of separation of the continents, and consisted of foundered continental crust (see Fig. 2.4), but the paleomagnetic dating of the ocean floor of the Atlantic and Indian oceans in the 1960s showed that the ridges were the most recently formed parts, and that the ocean floor became older towards the continental margins. This work proved that the continents of Gondwana and Laurasia had indeed moved apart, and that the space between had been filled by new ocean crust.

The dating of the ocean crust relies on the fact that new crust forming along the ocean ridges becomes imprinted with the contemporary magnetic field; this changes periodically because the Earth's magnetic field reverses at irregular intervals, each of several hundred thousand to several million years long, after which the magnetic north and south poles are swapped. Each of

these intervals creates a long strip of crust, parallel to the ridge axis, whose magnetic character differs from the previous one, and can be distinguished from it by remote measurement with a magnetometer. As new strips are created, older strips move away from the ridge axis. This process creates a series of magnetic bands or 'stripes' on the ocean floor, each of which represents a particular period of formation (Fig. 2.10). The stripe sequence was dated by comparing it with sequences of lava flows on land whose dates were known.

The age pattern of the Atlantic ocean floor given by palaeomagnetic measurements (Fig. 2.11) indicates a simple pattern where the youngest dates (0–10Ma) are in a central strip (red) along the axis of the Mid-Atlantic Ridge, proving that the ridge was the site of formation of new ocean crust at the present day. On each side of this central strip are successively older bands, with the oldest

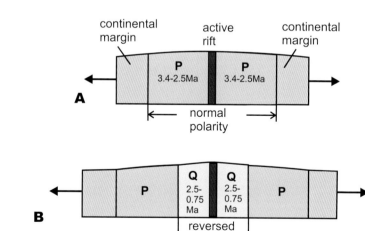

Figure 2.10 Oceanic magnetic stripe pattern. Three idealised cross-sections of ocean crust showing how stripes of alternating magnetic polarity are created every time there is a change from normal to reverse polarity: **A**, 3.4–2.5Ma ago; **B**, 2.5–0.75Ma; **C**, 0.75–Present.

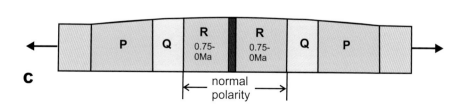

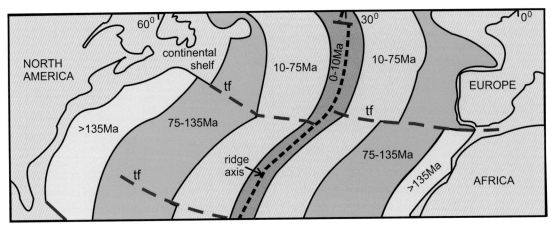

Figure 2.11 Dating the ocean floor. The age pattern of the Atlantic Ocean floor given by palaeomagnetic measurements indicates a simple pattern where the youngest dates (0–10Ma) are in a central strip (red) along the axis of the Mid-Atlantic Ridge. On each side of this central strip are successively older bands with the oldest (>135Ma) adjacent to the North American and African coasts respectively. These dates correspond to the dates when these continents originally separated. Note that north of the transform fault, the oldest dates are younger (75–135Ma) indicating that Europe separated from North America at a later date. tf, transform fault. After Larson & Pitman, 1972.

(>135Ma) adjacent to the North American and African coasts respectively. These dates thus correspond to the dates when these continents originally separated. Note that north of the fault separating Africa from Europe, the oldest dates are younger (75–135Ma) indicating that Europe separated from North America at a later date. (This fault is a transform fault, as explained in the following chapter.)

Ocean crust could not be continuously created without it being destroyed elsewhere (unless the Earth was expanding), and the obvious sites for its destruction were the deep-ocean trenches as proposed by Hess. The new palaeomagnetic dating evidence demonstrated that the ocean crust adjacent to the trenches showed a variety of ages (i.e. a 'discordant' age pattern) whereas the crust adjacent to continental margins that had moved apart showed a 'concordant' age pattern, as in Figure 2.10B, in which the oldest age was consistent with the date of separation of the continent. This evidence confirmed that the conveyor belt model for the ocean floor was essentially correct.

Now, thanks to the innovative genius of Alfred Wegener, Arthur Holmes and Harry Hess, married to the detailed and exhaustive field observations of Du Toit and the laboratory work of the geophysicists, all the preconditions for the plate-tectonic theory are in place.

3

Plate-tectonic framework

It is not possible to give a satisfactory explanation of the locations and origins of the Earth's great mountain ranges without some understanding of the processes that control the movements of the Earth's crust. The last chapter explored how ideas about these processes evolved gradually until, by the 1960s, it became generally accepted among geologists that the movements of the continents were controlled by lateral movements of the ocean crust, which themselves were propelled by solid-flow convection in the Earth's mantle. All that was needed was a conceptual model that explained how this process worked in detail – a model that came to be known as plate tectonics and which revolutionised the earth sciences because it provided an explanation for so much that had hitherto been unexplained or insufficiently understood.

The plate-tectonic theory evolved gradually out of the sea-floor spreading concept described in the previous chapter, thanks to the insights of several influential figures. Among these was Tuzo Wilson, of Toronto University, who recognised the significance of the linear zones of volcanoes and earthquakes that delineate a worldwide, tectonically active, network (Fig. 3.1). This map shows that the sites of the major earthquakes follow a network of narrow zones along the crests of the ocean

ridges and within broader and less well-defined zones along the island-arc–trench system of the Western Pacific and Indian oceans, and the orogenic belts of the continents (compare Fig. 2.8). The great majority of active volcanoes also occur within the same belts, although a few occur within the ocean basins, such as the Hawaian islands in the Pacific and the Cape Verde in the Atlantic.

Wilson realised that this concentration of tectonic activity must define the zones where crustal movements were taking place at present, and that the network of zones must therefore mark the boundaries of relatively stable 'blocks' of crust. The term 'block' was subsequently replaced by the now familiar term plate. The margins of these stable blocks could thus be traced by following the earthquake zones.

Transform faults

In a ground-breaking paper in 1965, Wilson pointed out that the numerous faults that appeared to offset the ridge axes and the trenches were also delineated by earthquakes, and were themselves part of the boundary of the stable blocks. He realised that these faults effectively changed the sense of movement across the boundary from, say, divergent across an

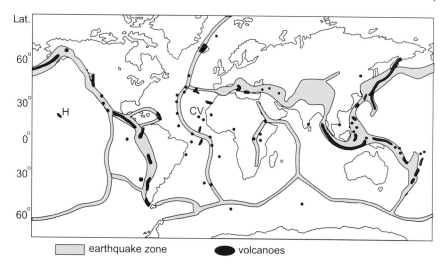

Figure 3.1 Global network of earthquakes and volcanoes. The earthquake zones follow the tectonically active crests of the ocean ridges, the island arcs of the western Pacific, and the Alpine–Himalayan and American Cordilleran mountain belts. The volcanoes are mostly concentrated along the same zones but some (e.g. Hawaii, H and Cape Verde, CV) also occur within the ocean basins. After Chadwick, 1962.

earthquake zone   ● volcanoes

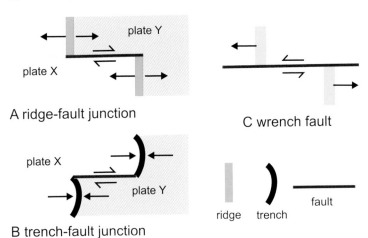

plate Y

plate X

A ridge-fault junction

plate X

plate Y

B trench-fault junction

C wrench fault

ridge   trench

fault

Figure 3.2 Transform faults. **A**, plates X and Y are moving apart from a ridge, and the sense of movement along the transform fault is dextral – i.e. the opposite side is moving to the right. **B**, plates X and Y are moving together at a trench; the sense of movement along the fault is also dextral. **C**, in a wrench fault, a green band on the opposite side of the fault is moved to the left and the sense of movement on the fault is the same as the movement of the green band – i.e. sinistral.

ocean ridge to one that was parallel to the fault. Such faults were given the name transform faults since they 'transformed' the sense of motion along the boundary of a block.

Figure 3.2 illustrates Wilson's concept with two examples: in A, plates X and Y are moving apart and divergent motion across the ridge is transformed to motion parallel to the fault that joins the ends of the ridge segments; in B, plates X and Y are moving together (Y is being thrust beneath X) and convergent motion across the trench is converted into motion parallel to the transform fault. Note that in both A and B, the sense of relative motion along the fault is dextral or right-lateral (i.e. for an observer standing on one side, the opposite side moves to the right). Figure 3.2C illustrates an example of a wrench fault (alternatively known as a strike-slip fault), which is a common type of fault found on land. This has apparently the same sense of displacement as the transform fault in A and B, in that the opposite side moves to the left. However here, the whole block on one side moves in the same direction and the sense of movement is sinistral, (or left-lateral). Thus an important property of a transform fault is that the apparent movement sense (in this case sinistral), had it been a wrench fault, is the opposite of the actual sense, which is dextral. So Wilson's discovery meant that many faults, including well-known examples of important faults that had previously been classed as wrench faults for many years, such as the San Andreas Fault of California, were re-classified by him as transform faults.

The North-East Pacific

The significance of Wilson's introduction of the transform fault is well illustrated by the North-eastern Pacific region (Fig. 3.3) Here, the western boundary

of the stable American Plate can be traced as an active earthquake zone from the Middle America Trench in the south to its intersection with the East Pacific Ridge, from where it continues north-westwards as the San Andreas Transform Fault until it meets the Cascadia Trench offshore from the Cascades volcanic arc. At the northwest end of this trench, the boundary continues along the coast as another transform fault, ending at the Aleutian Trench off Northwest Alaska. Each of these transform faults ends abruptly at a trench, at which point the sense of movement changes from one that is parallel to the fault to one that is convergent across the trench.

The Cascadia Trench sector exists because of a small oceanic plate (the Juan de Fuca Plate) situated offshore from the trench and separated from the main Pacific plate by a short section of ocean ridge, the Juan de Fuca Ridge. Within the Pacific Plate itself, all the transform faults are now inactive, the only currently active sections being those offsetting the active sections of the East Pacific and Juan de Fuca ridges. The inactive transform faults, which trend approximately E–W, are parallel to the former direction of relative motion between the Pacific Plate and the old Farallon Plate, now subducted beneath North America.

The Benioff–Wadati zones

The distribution of earthquake foci can also be used to demonstrate what happens to the stable blocks where they converge, as is the case around much of the Pacific Ocean. In 1949, two seismologists, Hugo Benioff of the California Institute of Technology and Kiyoo Wadati of the Japan Meteorological Agency, independently discovered that deep-focus earthquakes along parts of the volcanic island arc–trench network fell on a plane that dipped at an angle of around 45° beneath the line of the

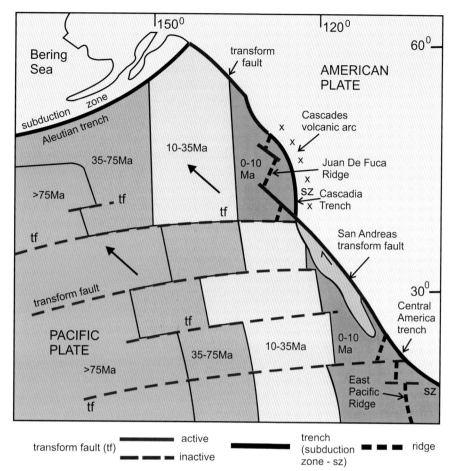

Figure 3.3 The NE Pacific. The boundary between the North American Plate and the Pacific Plate is defined by both subduction zones and transform faults (including the famous San Andreas Fault). In the Pacific plate the magnetic stripe pattern defines several dated zones which are offset by a series of transform faults. The East Pacific and Juan De Fuca ridges end against the San Andreas Fault and the movement direction changes from convergent across the subduction zones to margin-parallel along the fault. The transform faults within the Pacific are all parallel to each other and represent the direction of movement of the Pacific plate relative to the East Pacific ridge. In the north, the Pacific plate is being subducted beneath the Aleutian trench. Note that the movement direction of the Pacific plate relative to North America is NW-wards, parallel to the San Andreas Fault, and that the piece of North American continent west of that fault is also travelling NW-wards, relative to the rest of the continent. sz, subduction zone; tf, transform fault. After Larson & Pitman, 1972.

trench, and these zones became known as Benioff or Benioff–Wadati zones (Fig. 3.4) and subsequently as subduction zones. Detailed studies of the earthquakes in these zones were thus able to define not only the direction of movement but also the angle of inclination at the earthquake focus, and confirmed Hess's earlier suggestion of a downward movement of the oceanic crust along the trench network.

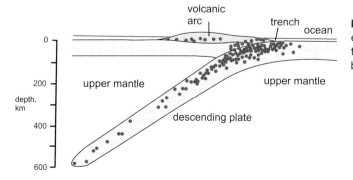

Figure 3.4 The Benioff–Wadati zone. The deep earthquakes (red dots) lie in a plane inclined beneath the volcanic island arc. Shallow earthquakes occur beneath the volcanic arc. After Uyeda, 1978.

The concept of the rigid plate

The final stage in the construction of the plate-tectonic model was based on the proposition that both continental and oceanic crust must behave in a semi-rigid manner, moving laterally as single units or blocks, and that relative movement between the blocks was concentrated at their boundaries. This insight arose mainly from the work of three scientists, Jason Morgan of Princeton, Dan McKenzie of Cambridge, and Xavier Le Pichon at Lamont Geological Observatory (now Lamont-Doherty of Columbia University), who all appear to have developed more or less the same concept independently.

The evidence for the rigidity of the blocks arose from the observation that linear features on the ocean floor, such as the faults and the striped magnetic pattern, were essentially unaffected by warping or bending such as might be expected if the ocean floor were to behave in a 'plastic' manner. The opposing coastlines of Africa and South-Central America still show a good fit despite having travelled away from each other for a distance of about 3000 kilometres over a period of 150 million years. Thus South America and the western half of the South Atlantic, on the one hand, and Africa and the eastern half of the South Atlantic on the other, could be considered as separate blocks, which moved as units. The Atlantic oceanic age pattern (see Fig. 2.10) confirms this.

The basic idea of the rigid plate is illustrated in Figure 3.5. Three blocks are shown: A, B and C (the term block was subsequently replaced by 'plate'). Plates A and B are moving apart away from a ridge, which is offset by two transform faults. Because the motion of the plate must be parallel to the fault, this determines the relative movement direction of A and B. On the left side of plate B, B is descending beneath C at a trench; this process is known as subduction and the trench marks the position of the subduction zone. Again, the movement direction of plate B relative to plate C must be parallel to the transform fault that offsets the subduction zone. This direction is oblique to the previous one – in other words, the

directions of relative movement of plate B are different for each boundary.

Thus an important property of plates is that they may move obliquely to a ridge or a trench but must always be parallel to any transform faults at their boundaries. At first sight, it may look as if plate B is moving in two different directions, but this is an illusion: both these movement arrows represent relative movements across different boundaries; this will be understood more easily if plate B is regarded as stationary; then plate A is moving away from it in one direction and plate C towards it in another direction. The principle that rigid plates can move obliquely to a convergent boundary was an important insight; it explained how mountain chains such as the Alps or Himalayas have curved, or even quite contorted, shapes yet are caused by the convergence of only two rigid plates whose movement directions must therefore be quite oblique to some sectors of the chains.

Lithosphere and asthenosphere

The rigid plates were conceived as having a definite thickness determined by the depth at which the material of the Earth's mantle ceases to behave in a semi-rigid manner and becomes capable of slow plastic flow in the solid state. The mantle within this zone of flow was termed the asthenosphere and the rigid plate above it, the lithosphere; the depth of the lithosphere varies, being thinner near the warm ridge and thicker at the cooler trench, but the mean thickness is around 100km. The lithosphere consists of the crust, which is the outermost layer of the solid Earth, plus part of the mantle, which lies beneath. It is this difference in properties between the lithosphere and the asthenosphere that enables the plates to move laterally over the weaker asthenosphere.

Tectonics on a sphere

Morgan (1968) had the idea of transferring the plate movements onto a spherical surface, having noted that

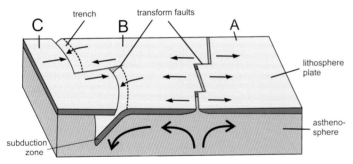

Figure 3.5 The lithosphere plate. Model showing three lithosphere plates, A, B and C. A and B are moving apart at a ridge offset by two transform faults. The movement direction is parallel to the transform faults. B is being subducted beneath C; here the movement direction is slightly oblique to the trench but parallel to the transform fault. After Isacks, Oliver & Sykes, 1968.

transform faults, especially in the oceans, were usually curved. He divided the Earth's crust into twenty 'blocks', all assumed to be perfectly rigid, whose relative motions could be described by rotations about a single axis emerging somewhere on the Earth's surface as a 'pole' (Fig. 3.6). MacKenzie, along with R.L. Parker (McKenzie and Parker, 1967), applied the same reasoning to the north-eastern Pacific Plate, which was described above (see Fig. 3.3), noting that the transform faults within the plate were all parallel and slightly curved. They showed that the motion of the Pacific Plate relative to the East Pacific Ridge was westwards, parallel to these transform faults, whereas its motion relative to the North American Plate was north-westwards, parallel to the San Andreas Fault, and towards the Aleutian subduction zone (i.e. in the direction of the arrow on Fig. 3.3).

Morgan noted that the transform faults in the Central Atlantic Ocean approximated to small circles whose centre was located near the southern tip of Greenland. He also discovered that the Atlantic Ocean spreading rates varied from a maximum in the central Atlantic to a minimum north of Iceland, and concluded that both these sets of data were consistent with motion about a single axis of rotation.

The plates

Le Pichon's contribution (Le Pichon, 1968) was to expand this data set over the whole ocean ridge network,

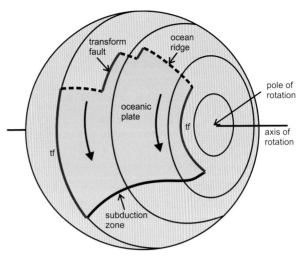

Figure 3.6 Plate tectonics on a sphere. The model illustrates the principle of plate tectonics on a spherical surface. The oceanic plate (green) is moving away from an ocean ridge and is being subducted beneath a trench. The movement direction is parallel to the transform faults that offset the ridge and also form the boundaries of the oceanic plate. The movement takes place around an axis of rotation that intersects the surface at the pole of rotation; the transform faults describe arcs of small circles about this pole. tf, transform fault. After Dewey, 1972.

which enabled him to simplify the global plate structure into just seven major plates: American, Eurasian, African, Indian, Australian, Antarctic and Pacific, together with a small number of minor ones (Fig. 3.7). He showed that

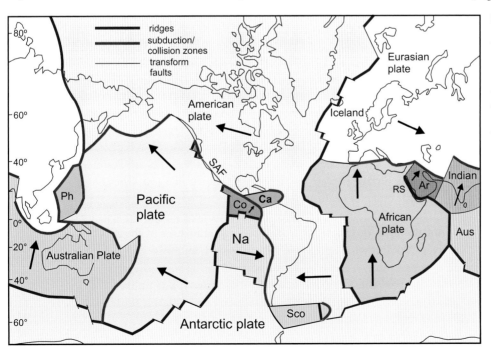

Figure 3.7 The plates and their boundaries. There are seven major plates: American, Eurasian, African, Indian, Australian, Pacific and Antarctic, together with several minor ones: Arabian (Ar), Cocos (Co), Caribbean (Ca), Nazca (Na), Philippines (Ph) and Scotia (Sco). The plates are separated by three types of boundary: constructive (ridges), destructive (trenches and collision belts) and conservative (transform faults). The arrows give the direction of motion of each plate relative to the Antarctic Plate (regarded as stationary). After Vine & Hess, 1970.

the movements responsible for opening all the major oceans – the Atlantic, Indian, Arctic, and the North and South Pacific – could each be explained by rotations about a single axis. The poles for these axes were found by the two independent methods: first by measuring the variation in spreading rates across the ocean ridges, as determined by the ages of the magnetic stripes, and secondly by plotting the intersections of the radii of the transform faults.

The spreading rates he measured varied from around 2cm per year (20mm/a) in the North Atlantic to more than 60mm/a in the central Pacific. By combining the data, it was possible to compute the convergence rates across the trenches: 60mm/a at the Peru Trench, 50–110mm/a across the trenches of the Western Pacific, and 56mm/a across the Himalayas.

Plate boundaries

The stable tectonic plates are completely surrounded by boundaries which, as we have seen, are of three types: ridges, trenches and faults. Since the ridges must mark the sites where new ocean crust is produced, they have become known as 'constructive' boundaries, and since the trenches mark the sites of destruction of ocean crust, they are termed 'destructive' boundaries. The transform faults correspond to zones where one plate merely slides past its neighbour without either creation or destruction of crust taking place. Consequently, because plate is 'conserved' at such boundaries, they are known as 'conservative' boundaries.

Making new plate: constructive boundaries

Figure 3.8A shows how the creation of new oceanic lithosphere takes place. Some of the new material is in the form of basaltic magma produced by melting of the upper mantle within the hot, low-density region beneath an ocean ridge. Part of this magma is injected into the crust in the form of intrusive bodies (e.g. dykes and sills), and some is poured out at the surface as lava flows. These rocks form the new oceanic crust; beneath it, new mantle lithosphere is formed by the addition of cooled material transferred from the asthenosphere. The ridge as a whole may be over 1000km wide and 2–3km above the surrounding ocean floor, but the earthquake and volcanic activity are concentrated in a narrow central rift about 100km across. The ridge is uplifted because of the presence of so much low-density material beneath it, and this results in a process

of stretching that enables the new material to rise into the crust. As this new material is added, earlier-formed lithosphere moves sideways, cools, and gradually sinks to the normal level of the deep-ocean floor.

The process of plate construction can be studied on land in Iceland, which is the only exposed part of the Mid-Atlantic Ridge. Here, an average spreading rate of c.20mm/a has been achieved by the injection of a swarm of dykes along the ridge axis accompanied by a number of volcanoes, two of which (Eyjafjallajökull and Grimsvotn) have been recently active, with dramatic consequences: the 2010 Eyjafjallajökull eruption produced an enormous dust cloud that grounded much of the European air-line travel for six days. Surface measurements at three of the main volcanic centres reveal areas of uplift between 40km and 50km in diameter, which are considered to be due to the accumulation of magma at the crust–mantle boundary.

Continental rifts

Figure 3.8B shows how new crust is created on the continents. An uplifted region forms, similar to an ocean ridge, with a central rift valley into which are poured the lavas that result from the melting taking place beneath. Magma is also injected into the crust within the stretched region. If this process continues, the continental crust will separate completely and oceanic crust will form in the gap created by the extensional forces, as shown in Figure 3.8C.

The Red Sea–Gulf of Aden–African rift system (Fig. 3.9) is an example of how a new constructive plate boundary may form in continental lithosphere. The three rifts join in what is known as a 'triple junction'; in two of the arms of the system – the Red Sea and the Gulf of Aden – new oceanic crust has formed along the centre of the rifts and joins up with the western Indian Ocean ridge. These new constructive boundaries define a separate Arabian plate, which is moving northeastwards away from Africa. Both these rifts are destined eventually to become oceans as Arabia and Africa move apart. In this way the continents of the Americas, Europe and Africa would have separated during the split-up of Pangaea.

The geological history of the African rift system is well documented; an initial doming of the surface was followed by a collapse of the rifts, accompanied by extensive vulcanicity, but unlike the other two rifts, there has been no significant separation.

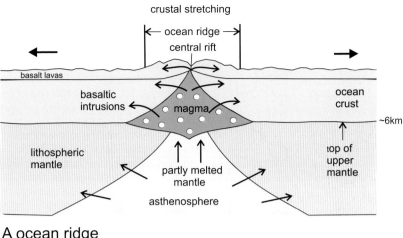

A ocean ridge

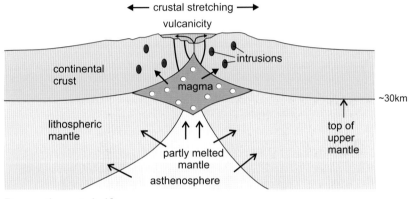

B continental rift

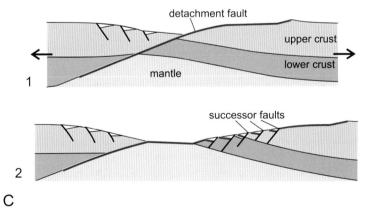

C

Figure 3.8 Constructive plate boundaries. **A**. Ocean ridge: new oceanic lithosphere is created by warm mantle material rising up into a region of stretched crust where it is partly melted; some of the resulting basaltic magma rises to the surface as lava flows and some is injected into the crust as intrusions. At the same time, the lithospheric mantle grows by the addition of cooled asthenosphere. **B**. Continental rift: the process is the same as A, except that the magma is emplaced within and onto continental crust. The stretched region forms a prominent rift valley. Eventually as the process continues, the continental crust separates completely, and oceanic crust forms in the rift, resulting in a new ocean. **C**. Extension in continental rift zones: 1, large amounts of extension cause initially steep faults to rotate due to solid-state flow in the lower crust, leading to arching up of the lower crustal material and the formation of a low-angle detachment fault; 2, further stretching and thinning causes the detachment fault to cut through the crust, exposing the uppermost mantle; successor faults may displace the upper part of the detachment fault. C, after Reston, 2007.

Destructive boundaries: subduction zones

There are three types of destructive plate boundary (Fig. 3.10). The first two types follow a zone of destruction of oceanic crust and are marked at the present day by the deep-ocean trenches; the third type is a zone of collision of two continental plates, and is represented by a belt of young mountain ranges such as the Alps and the Himalayas. This third type is, in geological terms, more short-lived, since the convergent movement of continental crust will eventually cease as the two continental plates grind together and gradually come to a halt.

The first two types of destructive boundary, therefore, are where oceanic crust belonging to one plate descends

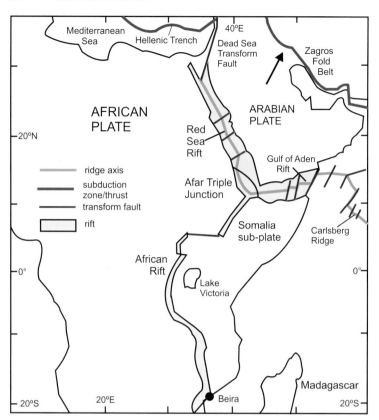

Figure 3.9 The Red Sea–Gulf of Aden–African Rift system. The Red Sea Rift extends from the Dead Sea Transform Fault to the northern coast of Ethiopia at the Afar Triple Junction, where it joins the Gulf of Aden Rift at the northern end of the great African Rift system. The Gulf of Aden Rift joins the NW end of the Carlsberg Ridge in the Indian Ocean to form a continuous boundary defining the western and southern margins of the Arabian Plate, which is moving NE-wards away from the African plate. Oceanic crust has formed in the Gulf of Aden and Red Sea rifts, which are moving apart. Extension has taken place across the African rifts, but no oceanic crust has formed there. The African and Gulf of Aden Rifts define the western margin of the small Somalia Sub-plate. After weebly.com/ somali plate, via Wikimedia Commons.

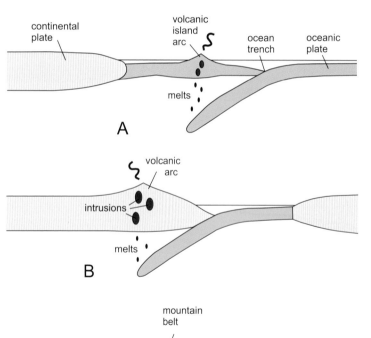

Figure 3.10 Destructive plate boundaries. **A** Oceanic subduction zone: oceanic lithosphere descends beneath another oceanic plate at a trench; melting occurs above the down-going slab, forming a volcanic arc on the upper plate. **B** Continent-margin subduction zone: oceanic lithosphere descends beneath the margin of a continental plate, forming a volcanic arc within the continental crust. **C** Continental collision zone: as two continents on opposing plates meet, the continent on the subducting plate is under-thrust beneath the continent on the upper plate; the continental crust is thickened and deformed and further convergence eventually stops. After Dewey & Bird, 1970.

(is subducted) beneath another plate, which may be either oceanic or continental (Figs 3.10A, B). The lower of the two plates is always oceanic, and the deep-ocean trenches mark the subduction zones where the oceanic plates commence their descent; such zones are typically inclined beneath the upper plate at an angle that can vary from quite shallow to nearly vertical, depending on several factors, such as the rate of convergence and the age of the descending plate.

As the lower plate descends into warmer regions of the mantle, volatiles, especially water, are driven out of the crustal material and ascend into the mantle of the upper plate, part of which melts; the resulting magmas form igneous intrusions within the crust of the upper plate and feed volcanoes at the surface. Where the upper plate is oceanic, as in Figure 3.10A, the volcanoes are partly submerged and form an island arc.

In a typical island arc, the partially submerged volcanic mountain range, which may be 50 to 100 kilometres wide, is bounded on its convex side by a trench situated between 50 and 150 kilometres from the volcanic arc. Between the arc and the trench is a zone where sediments resulting from the erosion of the volcanic islands accumulate, together with volcanic material. Present-day examples of volcanic island arcs are widely distributed in the western Pacific Ocean, the eastern Indian Ocean and in the Caribbean (see Fig. 3.1).

The sequence of marine sediments formed at a subduction zone is often given the generic term 'flysch' and typically includes a high proportion of turbidites; these are products of the chaotic mixing of clastic material originally deposited on the continental shelf but subsequently removed and redeposited on the continental slope or in the trench by mass-flow turbidity currents triggered by sudden instabilities such as earthquakes.

The continental-margin subduction zone

This type of subduction zone, which always precedes continent–continent collision, is represented in simplified form in Figure 3.11. The descending oceanic plate causes melting to take place in the mantle wedge above the subduction zone, leading to the formation of an elevated volcanic arc on the continent of the overlying plate. Sediments and volcanic debris derived from the erosion of this volcanic massif are deposited in the offshore ocean trench. Present-day continental-margin subduction zones are situated along the Pacific margins of South and Central America.

Chemical differences between magmas

The composition of the igneous rocks found within orogenic belts can be used to reconstruct the original tectonic setting in which they formed. Typical ocean-ridge basalts contain rather more silica than continental rift basalts, which are usually more alkaline. Subduction-zone magmas, although they include basalts, are typically more andesitic in composition and may also include more acid types such as rhyolite. Such magmas

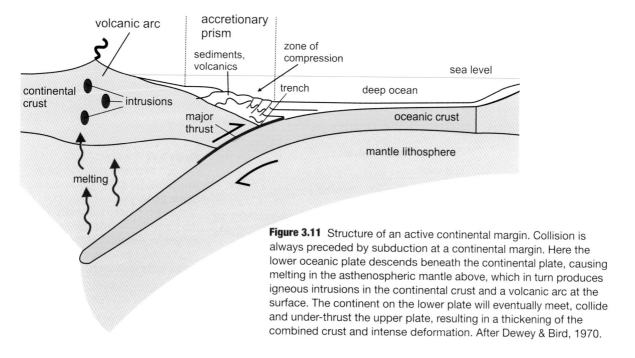

Figure 3.11 Structure of an active continental margin. Collision is always preceded by subduction at a continental margin. Here the lower oceanic plate descends beneath the continental plate, causing melting in the asthenospheric mantle above, which in turn produces igneous intrusions in the continental crust and a volcanic arc at the surface. The continent on the lower plate will eventually meet, collide and under-thrust the upper plate, resulting in a thickening of the combined crust and intense deformation. After Dewey & Bird, 1970.

are often referred to as calc-alkaline and typically result in the intrusion of granites or granodiorites. Significant differences are found in the minor (or 'trace') elements, and so measuring the relative proportions of these is an effective way of distinguishing between igneous rocks of differing origin.

The accretionary prism

The process of subduction generally results in an accumulation of sedimentary and volcanic deposits above the trench, caused by the wedge-shaped margin of the upper plate scraping off the cover of the down-going slab, and has been likened to the action of a snow-plough (Fig. 3.11). With time, the amount of material so formed increases and extends forwards onto the leading edge of the upper plate and also backwards along the top of the lower plate. This accumulation is known as an accretionary prism and includes platform cover, trench-fill deposits and material shed from the volcanic arc. The structures of the prism are typically in the form of a fold-thrust regime directed backwards along the lower plate, but may also feature forward-directed over-folds and thrusts on the upper-plate side.

Upper-plate extension

An important feature of many subduction zones, which may seem counter-intuitive, is a region of crustal extension on the upper plate, despite the fact that the two opposing plates are converging. The extension often leads to complete severance of the continental crust and the development of a small oceanic basin, called a back-arc basin behind the volcanic arc. The cause of back-arc extension is considered to lie in a process known as trench roll-back, where the position of the down-bend of the subducting slab (which determines the position of the trench) retreats backwards along the lower plate due to the downward pull of the cold, and thus heavy, slab (Fig. 3.12). Whether this takes place or not depends on a complex interplay of several factors, including the velocities of the opposing plates, the age of the descending slab, and the load of the accretionary complex on the upper slab. The process is influenced by the presence of a warmer upper mantle beneath the upper plate, caused by the release of hot material from the descending slab. There is evidence from seismic data to suggest that a piece or pieces of cold lithospheric mantle may become detached from the upper plate and sink, leaving the crust above warmer, weaker, and subject to extension.

Geological structure of mountain belts
Plate collision and mountain building

Where two continental plates collide, a zone of crustal overlap is created, which leads to a great increase in the thickness of the continental crust (see Fig. 3.10C). In some cases, crustal thicknesses of up to 80km have been recorded, e.g. in the Alps, compared with an average 'normal' crustal thickness of around 33km. Because continental crust is less dense and thus more buoyant than the underlying, denser, mantle, much of this extra crustal material becomes elevated to form mountain ranges with heights of up to 8km above sea level; due to the isostasy principle described in chapter 2, these mountain ranges are supported by a much greater thickness of crustal material beneath, forming a kind of mountain 'root'.

The thickened crust of the collision zone consists of rocks that have been intensely deformed by the effects of the collision: in places, great thrust sheets will have formed due to the sliding of the upper plate over the lower. As the crustal material becomes depressed into

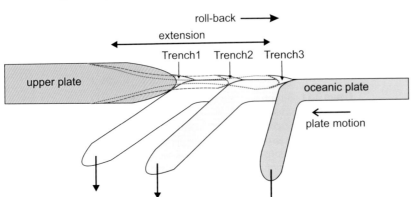

Figure 3.12 Trench roll-back. Cartoon to illustrate the roll-back mechanism. Assuming a stationary upper plate, the downward pull of the slab causes the position of the trench to migrate backwards, extending the upper plate. The dashed lines indicate the regions of upper-plate extension.

the warmer regions at depth, the folding and shearing processes caused by the collision are enhanced by the heating of the rocks caused by rising magmas. The effects of heat and pressure at depth produce changes in mineral composition and texture (i.e. metamorphism) of both sedimentary and igneous rocks.

Metamorphism

The type of metamorphism exhibited by the rocks of an orogenic belt may be used to distinguish between certain tectonic environments. Although both temperature and pressure increase with depth, their relative rates of increase differ according to the tectonic context. In the over-thickened warm crust of a collision zone, the temperature increase will have been steeper than the pressure increase, leading to a high temperature–low pressure series of metamorphic rocks ranging from greenschist facies to amphibolite facies and granulite facies as the temperature rises. In contrast, in the cooler crust of a subduction zone setting, pressure increases are dominant compared with temperature increases, resulting in a high pressure–low temperature metamorphic series typified by blueschist facies, then eclogite facies as the pressure rises.

Deformation

The style of deformation is another important indicator of the tectonic environment. Regions of the crust that have been subjected to compression exhibit contractional structures such as overthrusts and folds, which have resulted in a shortening and thickening of the crust. The upper or outer regions of a compressional regime are dominated by thrusts and the folds that are related to them, as seen, for example, in the outer zones of the Alps and the Scottish Highlands (e.g. see chapter 5, Fig. 5.5A and chapter 14, Fig. 14.8A) whereas at deeper levels where the temperature is higher, shortening is achieved by a combination of shear zones and ductile folds, as in the internal zones of the Alps and the Scottish Highlands (e.g. see chapter 5, Fig. 5.5B and chapter 14, Fig. 14.8B).

Extensional regions, on the other hand, exhibit a combination of normal faults and low-angle detachment faults (and their ductile equivalents at lower crustal levels), which result in the thinning and elongation of the crust (Fig. 3.8C). There is evidence in many mountain belts that extensional structures in the upper regions of the orogen were being formed at the same time as compressional structures were forming at the deeper levels, due to gravitational spreading of the over-thickened crust.

Foreland basins

The formation of a collisional orogenic belt places an additional weight on the lithosphere, resulting in a downward flexure that extends outwards beyond the margins of the mountain belt and produces basins on the foreland on each side of the belt. These basins gradually fill with sediment derived from the erosion of the mountain massif.

The sedimentary successions formed in this way generally show a transition from marine in the early stages to continental in the later, and are a useful indicator of the type of material being eroded at any one time from the uplifted core of the mountain belt. Sedimentary sequences of this type were termed 'molasse' by the early Alpine geologists and contrast with the flysch type of deposit typical of the accretionary prism in a subduction setting. A good example of the way that such basins develop and evolve with the growth of the orogenic belt is the Pyrenees belt, discussed in the following chapter (e.g. see Fig. 4.9).

An example of a collisional mountain belt

One of the best present-day examples of an active collision zone is provided by the mountain ranges of southern Asia (e.g. the Himalayan, Pamir and Tien Shan ranges) where the Indian continent has collided with, and underthrust, Asia (Fig. 3.13). The record of the gradual convergence of these two continents has been well documented from the ocean floor magnetic data (see chapter 8). Unlike constructive boundaries and transform faults, where the zones of earthquakes or volcanic activity are relatively narrow, in continental collision zones they can be over a thousand kilometres wide. The surface separating the two opposing plates may descend for long distances beneath the surface of the upper plate and may also be folded and faulted in a complex manner.

Figure 3.13 shows a wide zone of thrust slices consisting of sedimentary rocks that originally rested on the continental crust of the Indian plate, indicating that as this plate under-thrust Asia, the upper sedimentary cover, together with part of its basement, was pushed backwards and upwards to form the main Himalayan range. The overlying Asian plate, containing the remnants of the volcanic arc formed during the subduction stage, was also elevated to form the Tibetan Plateau.

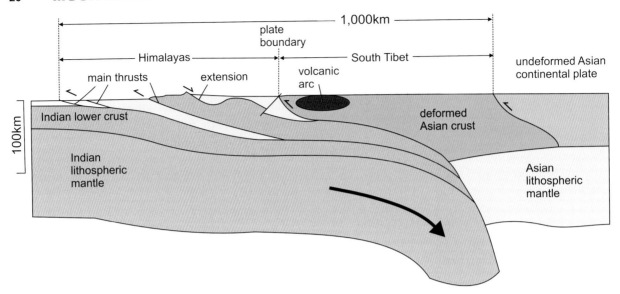

Figure 3.13 Structure of a collisional mountain belt: the Himalayas. The Indian plate (left) has under-thrust the Asian plate (right). Most of the deformation has occurred in the Indian crust, which has been sliced into several large sheets separated by major thrusts. The upper levels of the thickened crust have experienced extension. The leading edge of the Asian continental crust is much less deformed and contains the evidence of a volcanic arc produced by the previous subduction phase. The sedimentary cover of the Indian Plate is shown in yellow. After Searle *et al.*, 2011.

Typical structure of an orogenic belt

Several structural features that are found in most orogenic belts can be considered as typical: these are as follows.

1 The foreland, where the basement and sedimentary cover have not been affected by the orogenic deformation.

2 The foredeep basin, situated on the foreland but formed as part of the orogenic process, which has received sediments derived from the rising mountains of the orogenic belt; such sediments, typically non-marine, are often given the generic term 'molasse'.

3 An external zone where the strata are folded and thrust but have not been moved far from their original position – these are said to be autochthonous. Where only the sedimentary cover has been deformed but the underlying basement has not been involved, the deformation is said to be thin-skinned.

4 A zone of more intense folding and thrusting where several thrust slices contain material derived from the interior of the orogenic belt and are 'foreign' to their present environment – these are said to be allochthonous.

5 An interior zone containing rocks that have been brought up from a considerable depth and have been intensely deformed and metamorphosed under high temperature and pressure; this zone is often referred to as the metamorphic core complex, and represents the central over-thickened zone of the orogenic belt.

6 A marine basin containing sediments and volcanics originally deposited on the continental slope or in an ocean trench above a subduction zone; the sediments of this zone have traditionally been termed 'flysch' and the marine basin a 'flysch trough', by the Alpine geologists who first described them.

More complex examples

The main Himalayan range is a relatively simple example of a collisional belt, involving only two opposing plates. However, many other cases involve multiple collisions, creating much more complex mountain belts: some of the Alpine chains of the Mediterranean region discussed in chapters 5 and 6 are of this type. Figure 3.14 shows how a wider collisional belt may be produced by the collision of an island arc with a continent (stages 1 and 2) followed by a further collision with a second continent. Each collision would have been preceded by subduction, leading to the creation of a more complex geological record involving a number of events with different ages and different characteristics.

Interpreting past mountain belts

Mountain belts are usually referred to by geologists as orogenic belts (from the Greek *oros*, mountain, and *gen*, produce), and the processes responsible for

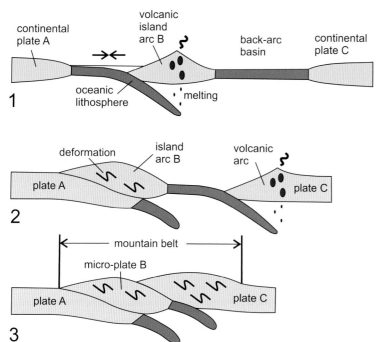

continental
plate A

volcanic
island
arc B

back-arc
basin

continental
plate C

1

oceanic
lithosphere

melting

deformation

island
arc B

volcanic
arc

plate A

plate C

2

mountain belt

micro-plate B

plate A

plate C

3

Figure 3.14 More complex collision belts. Many mountain belts are made up from the amalgamation of one or more island arcs with the main continental plates. Stage **1** shows subduction of an oceanic plate A beneath an island arc B, which is separated from continent C by a back-arc basin. In stage **2**, the island arc B has collided with plate A and subduction has commenced beneath plate C, leading to the subduction of the back-arc basin. In stage **3**, plate C has collided with the already-formed mountain belt created by the earlier collision to form an expanded mountain belt.

them are known collectively as orogeny (i.e. 'mountain-producing'). Thus a mountain belt formed in the far distant geological past, which may have no modern topographic expression but has been deduced from geological evidence only, will always be referred to as an 'orogenic belt'. The link between orogenic belts and plate collision is valuable in interpreting the geological record, especially in regions of old continental crust, formed in periods when no oceanic record has been preserved: all oceanic plates older than about 200Ma have been destroyed by subduction. Such belts display characteristic geological features, which can easily be identified in the rock record of former periods of Earth history.

These features include:

1 highly disturbed strata (i.e. folded and faulted rocks);

2 metamorphosed deep-crustal material (evidence of uplift);

3 great thicknesses of sediments derived from the erosion of uplifted masses; and

4 igneous rocks derived from melted lower-crustal material (e.g. granite).

4

The Western Mediterranean

The complex Alpine–Himalayan orogenic belt extends from the Atlantic Ocean at Gibraltar in the west to Myanmar (formerly Burma) on the eastern shores of the Indian Ocean (Fig. 4.1). It has been produced by the collision of three separate plates: the African, Arabian and Indian plates in the south, with the Eurasian Plate in the north. The shape of the belt is sinuous, and complex in detail, consisting of a large number of separate mountain chains, which in places enclose relatively flat, high plateaux, low-lying plains, and even ocean basins, such as those in the Western Mediterranean and the Black Sea.

There are three major mountain chains within the Western Mediterranean sector of the Alpine–Himalayan orogenic belt: the Pyrenees, the Betic Cordillera (or Betic Alps) in Europe, and the Atlas–Rif–Tell ranges in North Africa (Figs 4.2, 4.3). Two other less prominent ranges occur within the Iberian Peninsula: the Iberian Cordillera and the Catalan coastal ranges.

This western sector of the orogenic belt resulted from the convergence of Africa and Europe, which took place during a period of around 180Ma, beginning in the middle Jurassic Period, and culminating in the Miocene Epoch around 20Ma ago. Movement is still continuing, though at a much reduced rate of less than 10mm/a (per annum), and the area is subject to frequent earthquakes and vulcanicity. This essentially convergent movement was very complex in detail, and this partly accounts for the sinuous and complicated shape of the belt overall. Several crustal blocks (termed 'terranes') broke away from both the European and African margins and travelled independently as 'microplates' to form separate collisional units. Another complication was caused by the opening and spreading of several small oceanic basins within the orogenic belt, producing further distortions in its shape.

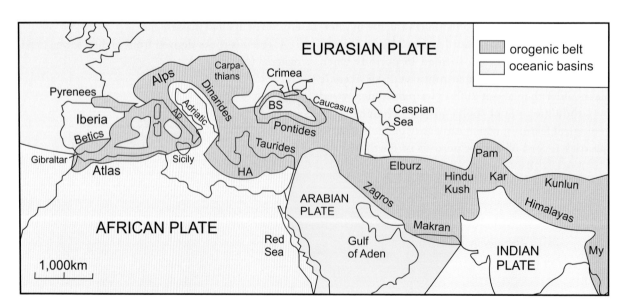

Figure 4.1 The Alpine–Himalayan orogenic belt. This belt is composed of many separate mountain chains that have been produced by collisions of three southern plates: African, Arabian and Indian, with the Eurasian plate in the north. In many places, the belt consists of northern and southern strands separated by lower-lying undeformed plains, or even ocean basins such as those in the western Mediterranean and the Black Sea. Ap, Apennines; BS, Black Sea; HA, Hellenic arc; Kar, Karakoram; My, Myanmar (Burma); Pam, Pamirs.

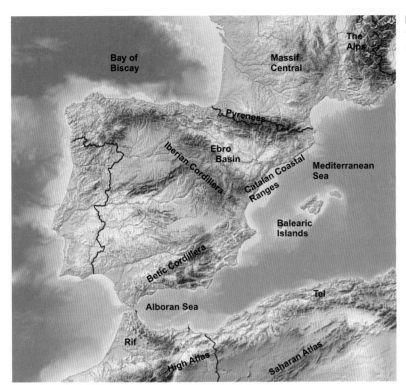

Figure 4.2 Mountain belts of the Western Mediterranean. Shutterstock©Schwabenblitz.

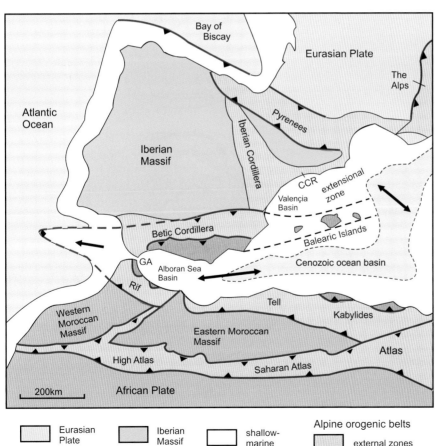

Figure 4.3 Alpine orogenic belts of the Western Mediterranean. GA, Gibraltar Arc; CCR, Catalan Coastal Ranges. After Sans de Galdeano, 2000.

Plate-tectonic context

200 million years ago, at the end of the Triassic Period, the Earth's surface would have appeared completely different to an observer viewing us from space. All, or almost all, of the continental crust was combined into a single supercontinent, Pangaea, which was composed of a northern grouping, Laurasia, and a southern, Gondwana. A worldwide ocean completely surrounded Pangaea and projected into it in the form of a large embayment known as the Tethys Ocean, which separated most of what is now Asia from the Arabian part of Gondwana (Fig. 4.4A). The floor of this ocean has completely disappeared, and parts of its oceanic crust, together with its sedimentary cover, have been incorporated into the various mountain chains. The present-day Mediterranean Sea, much of which is underlain by oceanic crust, is a more modern ocean, created during the Alpine orogeny, and known as the Neo-Tethys Ocean. The Alpine–Himalayan orogenic belt records the history of the break-up of Pangaea and the re-arrangement of its component parts into a different configuration.

Up until the middle of the Jurassic Period, the supercontinent of Pangaea was still intact, but during the later

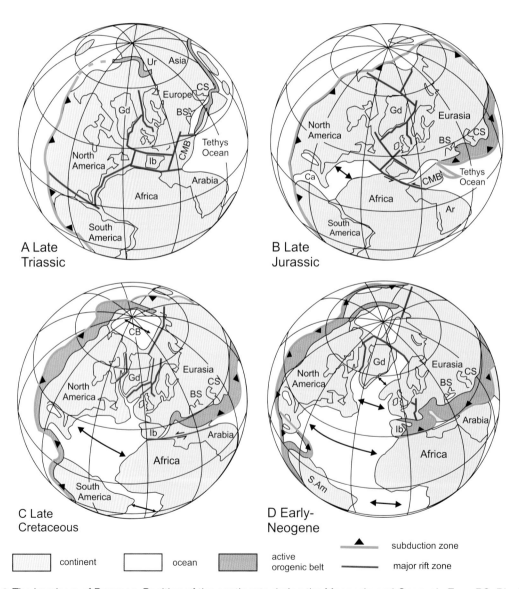

Figure 4.4 The break-up of Pangaea. Position of the continents during the Mesozoic and Cenozoic Eras. BS, Black Sea; Ca, Caribbean; CB, Canada Basin; CMB, Central Mediterranean Block; CS, Caspian Sea; Gd, Greenland; Ib, Iberia; S.Am, South America; Ur, Urals. After Ziegler, 1990.

part of the Jurassic, the Central Atlantic Ocean opened up due to a clockwise rotation of Laurasia relative to Africa, as shown in Figure 4.4B. This initial split in the supercontinent was bounded in the north by an arc-shaped transform fault running from Newfoundland through the Straits of Gibraltar, and in the south by a second transform fault along the north coast of South America.

In the mid-Cretaceous, the Atlantic Ocean spreading axis extended northwards into Eurasia, causing a change in the clockwise rotation of Eurasia (Fig. 4.4C). This resulted in Africa moving northeastwards and beginning to converge with southern Europe. It also caused the Bay of Biscay to open up, and Iberia to move southeast along the Bay of Biscay transform fault. Orogenic belts began to form along the various contacts.

During the Palaeogene Period, the Atlantic Ocean spreading axis moved further north to split Greenland from Norway, causing Eurasia to rotate further in a clockwise direction (Fig. 4.4D). Consequently, the convergence direction between Africa and Eurasia changed to a north-northeasterly and then a northerly direction. This caused collisional belts to be formed in the Pyrenees and along the whole length of an arcuate belt from Corsica, along the south-eastern coast of what is now France, into the Alps.

The relative movement path of Africa in relation to Eurasia has been accurately determined using palaeomagnetic ocean-floor data and is summarised in Figure 4.5; this shows a south-eastward motion through the later Jurassic and into the early Cretaceous, parallel to the Gibraltar transform fault. There was an abrupt change in the mid-Cretaceous to northeast, then

north-northeast, which persisted into the Palaeogene, when it became northwards until the Miocene Epoch of the early Neogene. Another abrupt change occurred near the end of the Miocene to northwest, which is the present-day convergence direction. Note that in this and succeeding chapters, compass directions are given relative to present-day co-ordinates.

Figure 4.6 summarises the movement history of the Western Mediterranean region during this period in simplified diagrammatic form. The changes are visualised by assuming that Europe (as part of the Eurasian Plate) was stationary and that all the relative movement was undertaken by Africa (as part of Gondwana) and the various terranes derived from it.

It is generally believed that Gondwana separated from the southern margin of Eurasia during the Jurassic by the opening of the Piémont and Liguria ocean basins (Fig. 4.6A). These were bounded to the southwest by the Gibraltar transform fault, which connected these movements to the opening of the Central Atlantic. During the early Cretaceous, a small terrane known as Alkapeca broke away from the Iberian margin, separated from it by a branch of the Ligurian Ocean. This terrane subsequently became separated into three parts, from which its name is derived: the Alborán complex in the Betics, the Kabylides, now in Algeria, and Peloritani–Calabria, now in SW Italy. Africa was then moving towards the southeast, relative to Western Europe.

In the mid-Cretaceous, the extension of the Atlantic Ocean spreading axis northwards into Eurasia resulted in Africa moving northeastwards and beginning to converge with southern Europe. It also caused the Bay of Biscay to open up, and Iberia to move southeast along

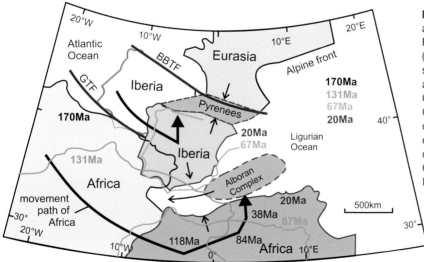

Figure 4.5 Movement paths of Iberia and Africa in relation to Eurasia. Positions from the mid-Jurassic (170Ma) to the Neogene (20Ma) showing outlines of the continents at two intermediate stages, early Cretaceous (131Ma) and end-Cretaceous (67Ma). Main areas of collision (Pyrenees and Alborán complex) shown in red. BBTF, Bay of Biscay Transform Fault; GTF, Gibraltar Transform Fault. Based on Handy *et al.*, 2010.

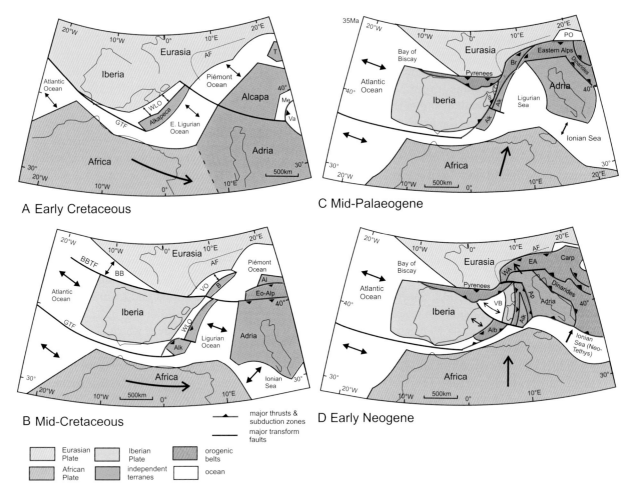

Figure 4.6 Plate tectonics of the Western Mediterranean. Interpretation of the sequence of movements of plates and terranes in the western and central Mediterranean from the beginning of the Cretaceous to the Early Neogene. AF, Alpine Front; Al, Alcapa; Alb, Alborán complex; Alk, Alkapeca; Ap, Apennines; B, Briançonnais Terrane; BB, Bay of Biscay; BBTF, Bay of Biscay Transform Fault; Br, Briançonnais Zone; Carp, Carpathians; EA, Eastern Alps; Eo-Alp, Eo-Alpine Orogen; GTF, Gibraltar Transform Fault; PO, Piémont Ocean; T, Tisza; Va, Vardar Ocean; VB, Valencia Basin; VO, Valais Ocean; WA, Western Alps; WLO, West Ligurian Ocean. Note that the Alpine belts of North Africa are not shown. After Handy *et al.*, 2010.

the Bay of Biscay Transform Fault (Fig. 4.6B). Three more terranes broke away from Africa during this period: Adria (the Adriatic Peninsula), Alcapa and Tisza; these are important in the evolution of the Alpine chains of the Central Mediterranean and will be discussed in the following chapter.

During the Palaeogene Period, the further extension of the Atlantic Ocean spreading axis between Greenland and Norway caused the convergence direction between Africa and Eurasia to change again, to a north-northeasterly direction (Fig. 4.6C). This caused collisional belts to be formed in the Pyrenees and along an arcuate belt from Corsica, along the southeastern coast of what is now France, into the Alps.

By the early Neogene Period, this convergent movement had changed to northwards, resulting in the formation of all the main mountain chains in the Western Mediterranean sector including the Alps, Apennines, Dinarides and Carpathians, partly as a result of the northwards movement of Adria and partly due to extensional movements in the Valencia Basin (Fig. 4.6D). The latter is a back-arc basin formed on the upper plate of a subduction zone situated along the eastern side of Alkapeca causing a block of crust consisting of Corsica–Sardinia and the northern part of Alkapeca to collide with Adria to form the Apennines.

The southern part of Alkapeca remained in the area now occupied by the Balearic Islands, and later in the

Neogene collided with Iberia in the north to form the Alborán complex and the Kabylides in Algeria in the south. During the later part of the Miocene, the Alborán complex moved south-westwards and was squeezed by the convergence between Iberia and Africa to form the Betic Cordillera.

The extensional movements in the Balearic and Alborán Sea basins are part of the eastwards migration of the subduction zone attributed to trench roll-back, and is discussed further in the next chapter.

The Pyrenees

There are four Alpine mountain belts in the Iberian Peninsula: the Pyrenees, the Betic Cordillera, the Iberian Cordillera and the Catalan Coastal Ranges (Figs 4.2, 4.3). Of these, the Pyrenean belt is the most dramatic and best studied.

The Pyrenean Mountains, named after the mythical Celtic Princess *Pyrene*, form a natural barrier between France and Spain, extending westwards from the Mediterranean coast at Cap de Creus, near Perpignan, to the Bay of Biscay at San Sebastián, a distance of 490km. However, the belt of Alpine deformation continues westwards, partly offshore, along the northern coast of Spain, to the western end of the Iberian Peninsula (see Fig. 4.7).

The main part of the belt, forming the Pyrenees, is between 100km and 150km across. The mountain range rises towards the east, and contains many high peaks including Pico de Aneto, the highest, at 3404m and Monte Perdido, 3355m. The highest summits are situated within the high mountain ridge of the Maladeta Massif, composed largely of granite (Fig. 4.8).

Tectonic setting

The Pyrenean belt originated in the Triassic Period as a major rift zone, the Bay of Biscay Rift, separating the Palaeozoic Iberian Massif from the rest of Europe (see Fig. 4.4A). This zone became a transform fault in the Jurassic Period, parallel to the important transform fault through the Straits of Gibraltar separating Iberia from Africa (Fig. 4.6B). During the Cretaceous, as the Atlantic opened further and spread north (Fig. 4.6C), the Bay of Biscay opened up between the marginal faults of the rift zone with the development of oceanic crust within the rift. This movement was associated with the counter-clockwise rotation of both Iberia and Africa relative to Eurasia.

Figure 4.6 shows how, between the mid-Jurassic and late Cretaceous, Iberia moved southeast parallel to the Bay of Biscay Transform Fault, which led to convergence and thrusting along the eastern segment of the belt. During the same period, Africa had moved by a much greater amount southeastwards, parallel to the Straits of Gibraltar Transform Fault, to a position close to its present one. By the end of the Palaeogene, the movement paths of both Africa and Iberia relative to Eurasia had changed to a more northerly direction, and the Bay of Biscay had begun to close due to convergence between Africa and Eurasia, causing compression along the length of the belt. Because the convergence is slightly oblique to the trend of the belt, the amount of tectonic shortening varies along its length, being much greater in the east.

The Pyrenees is a superficially symmetrical orogenic belt in the sense that it is bordered by forelands to both north and south, and thrusting is directed northwards in the north and southwards in the south, but it is asymmetric in detail, since the bulk of the thrust units are directed southwards. Five main tectonic units are recognised, from north to south: the Aquitaine Basin; the North Pyrenean Thrust Zone; the Axial Zone; the South Pyrenean Thrust Zone and the Ebro Basin (Fig. 4.7). The deep structure of the belt has been the subject of a detailed geophysical survey – part of a series termed ECORS (*étude continentale et oceanique par reflexion et refraction seismique*) conducted by an international team, which has resulted in the reconstruction shown in Figure 4.7B.

The nature and arrangement of tectonic units in the Pyrenees is inherited from its origin as an extensional rift zone (Fig. 4.7C); many of the major thrusts are 'normal faults' (i.e. they were formed by extension), which have been re-activated as 'reverse faults' (i.e. by compression) and the thrust sheets originated as extensional basins in the Cretaceous. This type of behaviour, where the movement on a fault reverses from extensional to compressional, is known as 'inversion' and is a common feature of orogenic belts. The predominance of southward-directed thrust sheets along the southern side of the orogenic belt reflects the original location of these basins south of the Europe–Iberia boundary.

The Aquitaine Basin

This basin encompasses a large area of southern France, bounded in the north by the Armorican Massif, in the east by the Massif Central, and in the south by the Pyrenean Mountains (Figs 4.2, 4.7). Most of this area was

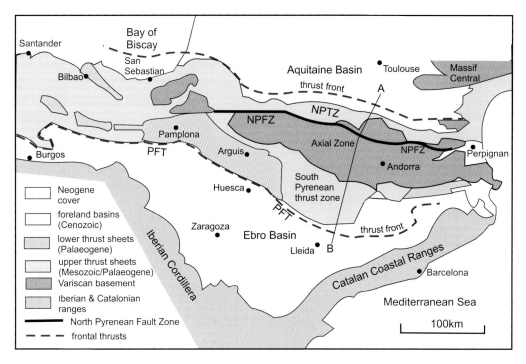

A

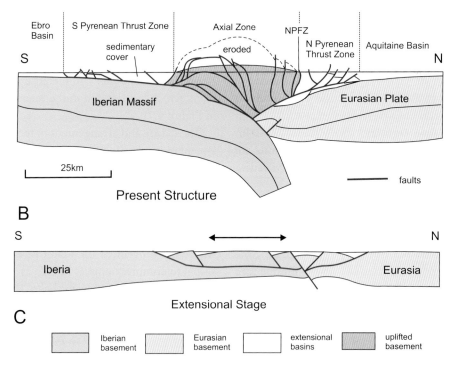

Figure 4.7 The Pyrenean Orogenic Belt. **A**. Simplified tectonic map showing the main tectonic zones. NPFZ, North Pyrenean Fault Zone; NPTZ, North Pyrenean Thrust Zone; PFT, Pyrenean Frontal Thrust. **B**. Simplified cross-section along the line A–B of Figure A to illustrate the crustal structure; note the complex antiformal thrust stack in the Axial Zone. **C**. Reconstruction of the late Mesozoic extensional stage; note (1) the series of half-grabens and (2) the crustal thinning resulting in a deep fault breaking through the crust; this will become the subsequent plate boundary. After Puigdefàbregas *et al.*, 1992.

Figure 4.8 The Maladeta Massif, Central Pyrenees. Taken from the Vall de Cregeuna, Posets. © Shutterstock, by patjo.

a continental platform throughout the Mesozoic and Cenozoic Eras and acquired a thin sedimentary cover commencing in the Triassic Period, which continued through the Jurassic. By the late Cretaceous Period, while typical shelf sedimentation continued in the north of the main basin, local small sedimentary fore-land basins developed close to what became the North Pyrenean Thrust, into which up to 11km of marine sediments were shed from the rising mountain range. Note that this is a cumulative figure obtained by adding thicknesses in several different places; no individual basin would be as deep as this. By the late Palaeogene Period, these basins had filled up, the sea retreated, and the Pyrenean fold-thrust belt began to advance across the foreland basins. Sedimentary sequences of this type are found throughout the Alpine–Himalayan belt and were recognised quite early in the history of geological investigation of the Alps as a type of tectonic unit integral to the orogenic process – thus 'synorogenic' – and the sediments characteristic of these units were named 'flysch'.

The North Pyrenean Thrust Zone

This zone is only about 10km wide over most of its length but broadens out to 40km in the west (Fig. 4.7). The central and western parts of the thrust zone contain outcrops of uplifted Palaeozoic basement, but most of the zone is composed of strongly folded Jurassic and Cretaceous platform sediments. By the later Cretaceous Period, an east–west extensional basin had developed within the zone, which received a thick sequence of flysch sediments. In the northern part of the zone, the Mesozoic sedimentary cover, including the flysch sequence, became detached from the basement uplifts and transported northwards on thrust sheets along a weak bed of Triassic evaporites (salt and gypsum). The thrust sheets extend across the southern margin of the Aquitaine Basin; the northern thrust front lies a short distance inside the Basin.

The southern part of the thrust zone, close to the North Pyrenean Fault, consists mainly of metamorphic rocks and contains many outcrops of ultramafic igneous rocks thought to have been derived from the upper mantle. The thrust zone is bounded on its southern side by the steeply-dipping North Pyrenean Fault Zone, which is regarded as the plate boundary between Iberia and the Eurasian Plate (Fig. 4.7). The zone becomes more diffuse north of Pamplona and its continuation westwards is uncertain.

The Axial Zone

This zone contains all the high mountains of the Pyrenean chain, including the Maladeta Massif, the central part of which is composed of a large granite intrusion. Much of the massif is over 3000m in height, and corresponds to a zone of thickened crust (Fig. 4.7B). This thickening has been achieved by a compressive shortening of approximately 20 per cent, equating to between 10km and 20km of horizontal displacement, but this reduces westwards as the zone becomes less pronounced. The zone extends from the Mediterranean coast south of Perpignan westwards for over 300km before being concealed by Mesozoic cover east of Pamplona. It is bounded on its northern side by the steeply-dipping reverse faults of the North Pyrenean Fault Zone, and on its southern side by steeply-dipping reverse faults that have back-thrust the overlying Mesozoic cover (Fig. 4.7B).

The rocks of the Axial Zone belong to the Palaeozoic basement, and are arranged in a series of thrust sheets. These have become arched up, and their boundary thrusts steepened due to the effect of the progressive forwards propagation of the basal thrust, as illustrated in Figure 4.9.

The South Pyrenean Thrust Zone

This is a southerly-directed fold-thrust belt, which is about 75km wide in the central part of the zone but decreases in width to both east and west (Fig. 4.7); its southern boundary is defined by the Pyrenean Frontal Thrust. There is considerable variation along the length of the zone: in the west, around Pamplona and Arguis, it is dominated at outcrop by thrust sheets consisting of deposits of Palaeogene age, originally laid down in a Cretaceous extensional basin like the one in the northern thrust zone. In the central part of the southern zone, thrust sheets containing mainly Mesozoic strata predominate, while in the eastern segment, there is a narrow Palaeogene basin in the northern part of the zone, but south of this there is a broad belt of thrust sheets containing platform sediments of the Ebro Basin, described below.

The earliest thrusts in the zone developed during the later Cretaceous Period by the reactivation of the normal faults bounding the early Cretaceous extensional basins. The thrusting then spread southwards during the Palaeogene Period until thrust movements ended around the end of the Palaeogene. At each successive stage in the deformation, as the thrust front migrated southwards, a depositional basin receiving molasse-type non-marine sediments formed in front of the rising mountains, but these were successively over-ridden as the thrusting progressed southwards. This process, illustrated in Figures 4.9A and B, was driven by the growth of the thrust stack and the consequent crustal thickening and elevation of the Axial Zone. Not all of the shortening was achieved by southward-directed thrusts; as compression at the margins of the Axial Zone increased, a northward-directed reverse fault (i.e. a 'back-thrust') developed along its southern boundary (Fig. 4.9). The thrust sheets rested on a bed of evaporite, which formed a weak layer over which they were transported.

The Ebro Basin

This basin occupies a roughly triangular area bounded on the north by the Pyrenean Frontal Thrust, in the southwest by the mountains of the Iberian Cordillera, and in the east by the Catalan Coastal Ranges (Figs 4.2, 4.9). The basin is highly asymmetric; the northern side is a foreland basin, and contains a thick sequence of clastic sediments. It is over-ridden by the South Pyrenean thrust sheets as described above; however, the remainder of the basin is an undeformed platform consisting of the Iberian Palaeozoic basement with a thin cover of Palaeogene and Neogene sediments. Marine deposits in the early- to mid-Palaeogene Period were succeeded by evaporites during the mid-Palaeogene, but from the late Palaeogene through the Neogene Period, the basin was surrounded by mountains from which it received non-marine (continental) clastic molasse deposits.

Minor Alpine mountain ranges of Iberia
The Iberian Cordillera

This rather subdued mountain chain extends from the western end of the Pyrenees in the north to the eastern end of the Betic Cordillera in the south (see Figs 4.2, 4.3), a distance of over 300km, and is about 250km across at its widest. It consists of part of the stable Iberian Massif, which has been affected by narrow zones of localised compression caused by thrusts and reverse

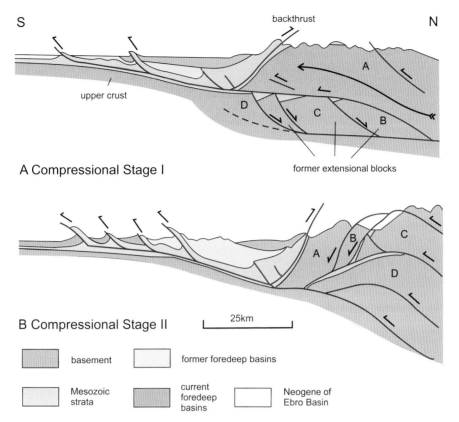

Figure 4.9 Evolution of thrusting in the Pyrenees. **A**. Compressional stage I (mid-Palaeogene): a basement block from the interior of the developing orogen has over-ridden the southern platform, pushing its cover before it; thrusts propagate southwards and where they break to the surface, elevations are formed, which shed clastic sediments into new foredeep basins. **B**. Compressional stage II (end-Palaeogene): a series of former extensional basement blocks have been successively overthrust towards the south, forming the antiformal thrust stack, thickening the crust and elevating the surface; the thrust front has propagated further into the foreland forming a series of new foredeep basins. The vertical scale has been exaggerated. The structure approximates to that of the cross-section A–B of Figure 4.7A. After Puigdefàbregas *et al.*, 1992.

faults, separated by wide zones of relatively undisturbed stable platform. The origin of the Iberian Cordillera probably lay in the set of extensional rifts and normal faults that existed during the Jurassic Period. The mid-Palaeogene convergent movement that was responsible for the Pyrenees also caused compression across the Iberian Cordillera by inverting the movements on the pre-existing normal faults.

The Catalan Coastal Ranges

This narrow mountain belt, only about 50km across, extends along the southeastern coast of the Iberian Peninsula, and forms the southeastern margin of the Ebro Basin from the Iberian Cordillera in the southwest to the Pyrenees in the northeast (Fig. 4.7). It was formed by a series of reverse faults and associated folds, similar to those of the Iberian Cordillera, but which trend NE–SW, at a high angle to both the Iberian and Pyrenean structures. The Catalonian structures appear

to be later than the others, and formed at a time when the convergence direction between Africa and Iberia had rotated into a more northerly orientation in the late Palaeogene. By the early Neogene, this coastal belt was affected by crustal extension related to the opening of the Valencia Basin, as discussed below (see Fig. 4.6D).

The Betic Cordillera and the Rif–Tell belt

The Betic Cordillera (Figs 4.2, 4.10) consists of a number of separate mountain ranges, situated along the southern coast of Spain, at the western end of the Mediterranean, between western Andalucia and Valencia. The cordillera as a whole is 500km long and up to 150km across, and is bounded in the north by the basin of the Guadalquivir River. The most impressive of the ranges is the Sierra Nevada (Fig. 4.11), which contains mainland Spain's highest mountain, Mulhacem, at 3478m, situated in the Sierra Nevada National Park.

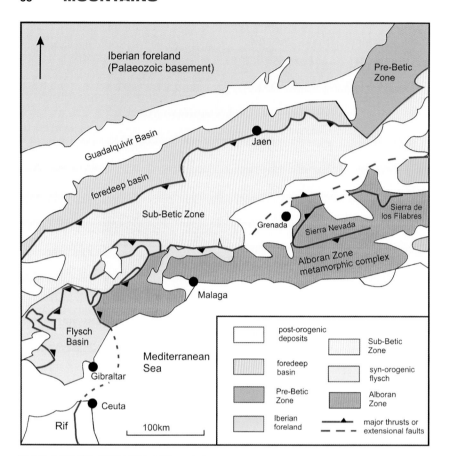

Figure 4.10 The western Betic Cordillera. After Alonso-Chaves *et al.*, 2004; and Castro *et al.*, 2009.

Figure 4.11 The Sierra Nevada from Guadix. © Shutterstock, by Arena Photo UK.

At its western extremity, the mountain belt turns through an angle of 150º around the Straits of Gibraltar into the Rif–Tell belt of Morocco. The eastern end of the chain is submerged beneath the Mediterranean Sea but reappears in the Balearic Islands. The Betic Cordillera is divided into three main tectonic domains: the External Region, the Flysch Basin and the Alborán, or internal, Zone.

The External Region

The External Region corresponds to the passive continental margin of the Iberian Peninsula, in the south-western part of the Eurasian Plate, and is equivalent to the foreland of the Alpine orogenic belt. It consists of a basement of deformed and metamorphosed Palaeozoic rocks with an unconformable cover of Mesozoic sediments typical of the continental platform. This sedimentary cover thins southwards towards the original passive continental margin and now forms a typical foreland fold-thrust belt in which the Mesozoic strata have been overthrust towards the northwest in a number of thrust sheets. Along the northern margin of the external zone, the Mesozoic platform sequence is overlain by Neogene sediments of the Guadalquivir Basin, the southeastern part of which consists of a foredeep basin, containing molasse deposits shed from the rising orogenic belt.

The main part of the External Region, south of the thrust front, has been subdivided into the Pre-Betic, Intermediate and Sub-Betic zones (Figs 4.10, 4.12). The Pre-Betic Zone consists of lightly deformed Triassic to early Neogene shallow-marine carbonate deposits of the foreland, which have not been displaced far from

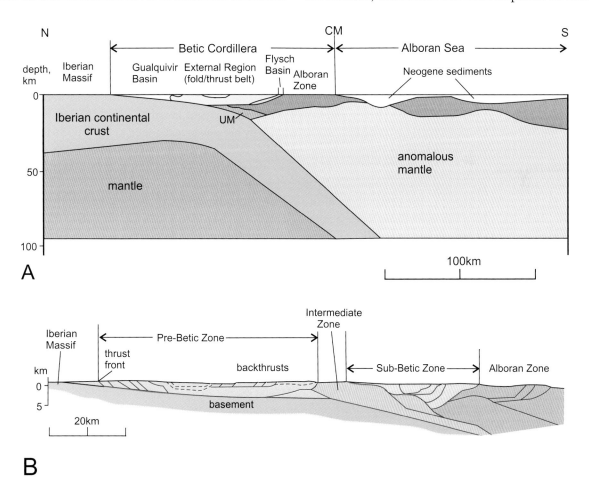

Figure 4.12 Large-scale structure of the Betic orogen. **A**. Interpretation based on geophysical data. Note: 1, the Iberian Plate descends southwards beneath the Alborán Sea; 2, the thinned and detached continental crust on the upper plate. CM, continental margin; UM, ultramafic body. **B**. Simplified NW–SE cross-section through the eastern part of the Cordillera, across the Sierra de Segura to the coast near Cartagena, showing the main tectonic zones and principal structures. Some faults are shown in red. A, after Azañón *et al.*, 2002 and Morales *et al.*, 1999; B, after Banks & Warburton, 1991.

their original site of deposition; such tectonic units are known as 'autochthonous'. However, in the Sub-Betic Zone, the Mesozoic sequence has been thrust northwards over the Pre-Betic Zone, and has lost contact with its original basement; such thrust units are known as 'allochthonous'. This zone occupies a wide area between the thrust front and the margin of the Alborán Zone in the area represented by Figure 4.12, having completely over-ridden the underlying zones in the southwestern and central sectors of the External Domain. The Intermediate Zone, situated between the Pre-Betic and Sub-Betic zones, contains material that originated in a late Jurassic to early Cretaceous sedimentary basin; it is not exposed in the area of Figure 4.10 but occupies a large part of the cross-section of Figure 4.12B.

The Flysch Basin

This is a complex of thrust slices containing clastic sediments of Cretaceous to Paleogene age deposited in a marine basin. It only appears at the surface in the southwestern part of the belt, northwest of Malaga and around Gibraltar (Fig. 4.10). The base of the flysch sequence in the Rif belt rests on basalt volcanics of the oceanic crust.

The Alborán Zone

This tectonic zone represents the metamorphic core complex, originally situated in the central part of the orogenic belt. It has been subdivided into three tectonic complexes, each of which contains a number of thrust slices: the Nevado-Filabride, Alpujarride and Malaguide complexes. These units have been arched up into a large, elongate dome structure that exposes the structurally lowest complex, the Nevado-Filabride, in its core. It is this unit that forms the high mountains of the Sierra Nevada within the Sierra Nevada National Park (Fig. 4.11) and also the Sierra Filabride to the east. It consists of metamorphosed rocks, mainly schists and gneisses, belonging to the Palaeozoic basement together with a thin Mesozoic to early Palaeogene cover.

The Nevado-Filabride Complex is flanked on its north and south sides by the structurally higher thrust sheets of the Alpujarride Complex, which consists of Palaeozoic and Permo-Triassic metasediments, including prominent marbles, and also contains thrust slices of peridotite derived from the upper mantle.

The structurally highest thrust units of the zone belong to the Malaguide Complex, exposed around the city of Malaga, and contain metamorphosed Palaeozoic basement rocks together with unmetamorphosed Mesozoic to Palaeogene carbonates and Neogene clastic deposits.

The boundaries between the thrust complexes of the Alborán Zone and between many of the individual thrust units within each of the complexes are now represented by extensional shear zones or faults, many of which originated as thrusts. Some of these extensional faults in the central part of the orogen are overlain by Neogene sedimentary basins, implying that the extensional movements also took place during the Palaeogene or early Neogene Periods and may have been partly contemporaneous with the compressional movements in the external zone.

The present complex arrangement of the Alborán structures is due to phases of N–S to NW–SE movements involving both compression and extension.

Tectonic history

The early tectonic history of the Betic region was dominated by extension in the Jurassic Period, which resulted in N–S to NE–SW-oriented rifts. These became filled with sediments of mostly Cretaceous age, and subsequently became incorporated in the Intermediate Zone of the fold-thrust belt.

The structural evidence from the thrust structures in the Alborán Domain indicates that these tectonic units were initially emplaced westwards, from the original interior of the orogenic belt that then lay to the east, in an area presently lying within the Western Mediterranean Sea (see Fig. 4.6D). The structures associated with the Alpine metamorphism of the Alborán rocks have been dated as early Miocene, but an older, Eocene, metamorphic event is also recorded, which must indicate the date at which the lower units of the complex became buried to some depth.

The main compressional event resulting in the foreland fold-thrust belt of the External Domain took place during the early Miocene Epoch, around 20Ma ago, and was a consequence of the change in convergence direction between Africa and Iberia shown in Figure 4.6. The thrust and fold structures of the Flysch, Pre-Betic and Sub-Betic zones indicate that these movements were directed towards the north and northwest in the outer parts of the fold-thrust belt, but also that backthrust movements in the inner part of the belt were directed towards the interior of the belt, as shown in Figure 4.12B. There has also been significant extensional faulting, especially in the Alborán Domain, which seems to have been, at least in part, contemporaneous with the compressional movements in the fold-thrust belt.

The early Miocene structures are affected by folding and reverse (i.e. compressional) faulting of late Miocene age, indicating a phase of NW–SE compression that was also responsible for the arching up of the Alborán units.

The Gibraltar Arc and the Rif–Tell belt

The Betic belt continues around the Gibraltar Arc, curving through 150° into the Rif–Tell belt of North Africa (see Fig. 4.3). The latter belt is a mirror image of the Betic Cordillera: the metamorphic core complex, continuation of the Alborán Domain, is exposed along the coast, but most of the belt lies within the External Region, where the structures are directed outwards towards the south. The Rif chain, lying entirely within Morocco, curves in an arc along the shore of the Mediterranean Sea from Tangier, at the Straits of Gibraltar in the west, to the Algerian border in the east, a distance of about 290km. Compared to the more impressive Atlas Mountains to the south, the Rif is not particularly high – the highest peak being Monte Tidirhine, at 2456m. The belt continues further east as the Tell range, which extends for 1500km along the northern coast of Algeria, finally merging with the Atlas belt in northern Tunisia. The western closure of the Gibraltar Arc is hidden beneath the sea but can be clearly traced as a submerged ridge, between 300m and 900m in depth, linking Gibraltar in the north with Ceuta in the south.

Tectonic interpretation of the Betic–Rif–Tell belt

The Pyrenees belt, which was discussed earlier, has a relatively simple overall structure caused by extension followed by compression – both directed roughly at right angles to the trend of the fold belt. In comparison, the Betic Cordillera is much more complex, and it is clear that for orogens of this type, the plate-tectonic model offers only a partial explanation. Two other processes have to be considered in explaining them: the effects of gravitational imbalances caused by the over-thickening of the crust, and the relative movements of blocks of crust that are smaller in scale than the main plates, but which interact with them. This latter process will become evident in considering the Alps.

During the early stages of the Betic–Rif–Tell history, the contiguous regions of Africa, Iberia and North America were part of a continuous continental platform, whose passive margin lay far to the east at the western margin of the Tethys Ocean (see Fig. 4.4A). During the

Triassic and Jurassic Periods, this platform was cut by a number of extensional rifts, which became filled with sediments during the Jurassic and Cretaceous Periods, and some of which expanded to become ocean basins. One of these oceanic basins became what is known as the Piémont–Liguria Ocean, which lay east of Iberia, roughly in the position now occupied by the Balearic Islands; another became the mid-Atlantic Ocean. At this time the southern margin of Iberia was a transform fault that ran through the Straits of Gibraltar as shown in Figure 4.6.

The tectonic situation changed in the Paleogene when the African Plate began to converge with Eurasia (see Fig. 4.6C, D). Compression caused by this change in relative motion resulted in the stacking of thrust slices, giving rise to extreme crustal thickening and the development of high-pressure metamorphism in the lowest units of the Alborán Domain, which at that time formed part of the Iberian platform lying east of the present southeast coast of Spain.

As the convergence between Africa and Iberia continued through the Neogene, the Alborán complexes were emplaced onto the Iberian platform and the Iberian lithosphere was under-thrust beneath the Alborán units, resulting in the development of the fold-thrust belt of the External Region. At the same time, the Rif–Tell belt was forming along the northern margin of Africa as the Alborán Zone was squeezed by the convergence of Africa and Iberia, resulting in the internal units being emplaced southwards on the opposite side of the orogen.

The original tectonic sequence has been disrupted by extensional movements, which have resulted in gravitational sliding of the uppermost units of the thrust stack. These seem to have been related to later Neogene N–S compression, which resulted in the arching up and exposure of the lower units of the Nevado-Filabride Complex.

It has been estimated that about 250km of north–south crustal shortening took place across the Betic belt during the late Palaeogene to early Neogene phase of collision between the African Plate and Iberia. This was followed by around 50km of oblique convergence in a NW–SE direction in the later Neogene.

Considering the Betic–Rif–Tell belt as a whole, the most obviously strange feature is the gap in the centre corresponding to the Alborán Sea. The section in Figure 4.12A shows that this is partly due to a thinning of the continental crust and partly to complete rifting leading to the development of a section of oceanic crust. So what should have been the thickest and most elevated sector of the orogenic belt has collapsed. These movements also resulted in pronounced extension within the Betic belt. There have been differing views as to the reason for this Neogene extension, but it is generally attributed to processes taking place within the mantle, and correlated with the gradual withdrawal eastwards of the subduction zone at the margin of the Neo-Tethys Ocean, discussed further in chapter 5. According to this interpretation, the Alborán Sea extensional region is a back-arc basin situated on the upper plate of a subduction zone that is retreating eastwards due to the process of 'slab roll-back' (see Fig. 3.12). Because the Alborán Basin represents an early stage of this process, there is little direct evidence of it here, but further east, in the Calabrian Arc where the subduction zone is situated now, there is clear evidence, which will be discussed in chapter 5.

During the later Miocene, the Alborán core complex was squeezed westwards into the Atlantic to form the Gibraltar Arc, and eastwards into the Mediterranean by the continuing convergence between the African and Eurasian plates (Fig. 4.13).

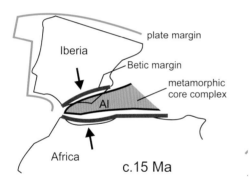

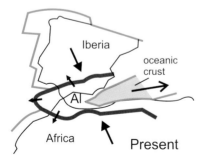

Figure 4.13 The Gibraltar Arc. Neogene NNW–SSE convergence between Iberia and Africa caused crustal thickening of the Alborán core complex, resulting in outward-directed thrusting around its perimeter (the Gibraltar Arc) followed by extensional collapse and eastwards escape of the central part. Al, Alborán region.

The Atlas Mountains

The Atlas Mountains consist of several distinct mountain chains stretching from the Atlantic coast of Morocco to the Mediterranean coast in Northern Tunisia, a distance of 2500km (Figs 4.2, 4.3). The mountain belt forms a barrier to the Sahara Desert to the south. The westernmost chain, 700km long, is known as the High Atlas and contains the highest peak in the range, Jebel Toubkal 4167m (Fig. 4.14). East of this is the Saharan Atlas, which extends for a further 1800km through Algeria in a northeasterly direction towards the Tunisian border where it meets the Tell Atlas.

Unlike the Betic–Rif–Tell belt, the Atlas chains are an orogenic belt that is entirely intra-continental in origin; there is no record of the subduction of oceanic crust, and the mountain chains have resulted from the compression and uplift of what was an extensional rift basin formed during the Mesozoic Era. The extensional faults that formed the boundaries of the rift valley became re-activated as reverse faults or thrusts during the late Palaeogene to early Neogene compression caused by the convergence between the African and Eurasian Plates. The belt as a whole is symmetrical, in the sense that the thrusts and overfolds at the margins of the belt are directed outwards – to the north along the northern boundary and to the south along the southern.

The structural evidence from the faults and folds along the marginal zones of the orogenic belt indicates that the direction of compression was oblique, causing a mixture of thrusts with a north–south movement direction and dextral (right-lateral) strike-slip faults. This type of arrangement is known as 'strain partitioning' and indicates that the convergence direction, at least in its latter stages, must have been oblique to the trend of the fold belt – in this case NW–SE rather than north–south, similar to the later convergence direction recorded in the case of the Betic belt.

Figure 4.14 The High Atlas range. Jebel Toubkal National Park, Morocco. Shutterstock©John Copland.

5

The Central Mediterranean: Alps and Apennines

East of the Iberian Peninsula, the shape of the Alpine–Himalayan belt in the Central Mediterranean is extremely complex: from southeastern France, the northern branch, or European Alps, curves around through nearly 180º, describing a great arc through eastern France and Switzerland until, in the Austrian sector, it trends roughly east–west (Fig. 5.1). The chain then divides into two branches, one continuing down through the Dinarides to Greece, and the other forming the great arc of the Carpathians. Another branch continues from the west end of the Alpine chain along the western side of Italy to form the Apennines.

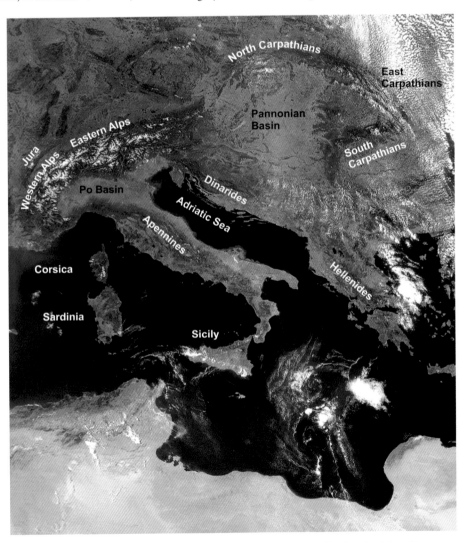

Figure 5.1 Mountain belts of the Central Mediterranean. Shutterstock©pio3.

The Alps

To most Europeans, and certainly most mountaineers, skiers and geologists, 'the Alps' means the mountain chain stretching from southeastern France, through Switzerland, Italy and Austria; however, the term 'alps' is also applied to other mountain chains such as the Betic Alps of southern Spain and the New Zealand Alps. The word 'alp' itself is a Celtic word applied to any high mountain area (hence the old word 'Alba' for Scotland) and is also used by the Swiss to mean an area of high pasture!

This chapter deals with the sector of the Alpine–Himalayan belt that extends from the Mediterranean coast at the French–Italian border near Monaco to just south of Vienna in eastern Austria, a distance of approximately 800km. This chain includes many famous summits, including Mont Blanc, the Matterhorn, and the Eiger, which have attracted mountaineers since the earliest days of the sport, and are among around a hundred peaks over 4000m in height. Geologists generally divide the sector into the Western Alps of France and Switzerland, the Eastern or Austrian Alps, and the Southern or Italian Alps. Most of the higher peaks are in the western sector. There are distinct geological differences between the three sectors, as will be seen.

Early geological investigations

The Alpine chain was not only the first of the great mountain belts to be studied by geologists, but also the one most intensively investigated subsequently. Most of the early geologists, including both James Hutton and Charles Lyell (see chapter 2) visited the Alps and recognised that the geological structures found there signified that lateral compression as well as uplift had played a part in their development. One of the most important contributions was made in 1837 by the naturalist Louis Agassiz (Agassiz, 1840), who was the first

to realise that the Alps must have been covered by ice to a much greater extent than at present, and that many of the erosional features of the mountains were due to the effects of glacial erosion, including the characteristic shapes of glacial valleys and the importance of glacial deposits such as moraines and glacial erratics. Agassiz subsequently visited Scotland and demonstrated the same features there, establishing the former existence of a great northern ice sheet covering much of Britain.

By the third decade of the twentieth century, the broad stratigraphy of the Alps was well known and its structural complexity widely recognised. Detailed studies of the stratigraphy and structure had revealed the presence of giant thrust sheets, termed 'nappes', enclosing complex recumbent folds, which had resulted in the same stratigraphic sequence being repeated upwards several times, accounting for considerable crustal thickening. A number of well-known geologists, including Emile Argand (1916), Albert Heim (1921), Leon Collet (1927) and Rudolf Staub (1928), had produced syntheses of Alpine geology, but the most important early attempt to understand the overall structure was made by Argand in his 1916 work Sûr l'arc des Alpes occidentales (Fig. 5.2).

The exponents of the continental drift theory believed that the overall structure of the Alps resulted from a collision between Africa and Eurasia and that Africa, or part of Africa, had effectively over-ridden Europe. In his influential 1937 work Our Wandering Continents, referred to in chapter 2, Du Toit visualises the sequence of events affecting the Alpine region as follows.

1 Latest Cretaceous Period. Eurasia and Africa began moving towards each other, creating 'geanticlinal' systems (i.e. uplifts) along their opposing margins.
2 Later Eocene Epoch. The Tethys Ocean became narrower and Western Europe became weakened by the 'break-away' of North America.

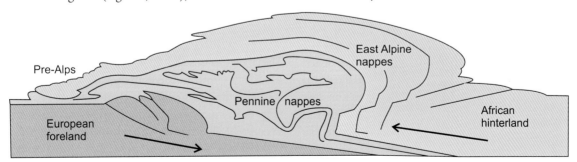

Figure 5.2 General section across the Alps according to Argand. Note that material from the African hinterland is thrust over the Pennine nappes and that the complex folding of the Pennine nappes directed both to the northwest and southeast is explained by the convergent movement of the African 'indenter'. After Argand, 1916.

3 Oligocene Epoch. General northwards advance of Africa resulted in squeezing of part of Africa north-westwards into the arc of the Western Alps, creating an 'S-shaped' line of collision.

4 Pliocene Epoch. Further north–south convergence resulted in a pressing together of the arms of the 'S' producing overfolding both to the north and south.

In terms of the timing of events, this account is not dissimilar to the more modern versions discussed below; however, it should be noted that Du Toit and other contemporary exponents of the 'mobilistic' view of tectonics believed that the continental crust could behave in a much more 'plastic' manner than subsequent 'rigid' plate theory demanded.

Plate-tectonic context

One of the problems facing those attempting to explain Alpine structure by conventional plate-tectonic theory was how structures that appeared to imply movements directed north, northwest, west and southwest could be produced by the convergent motion of two rigid plates. It became clear that the overall structure of the Alps could only be explained by the relative movements of several independent terranes, as described in the previous chapter, together with changes in the convergence direction over time. Those movements were summarised in Figures 4.4 and 4.6.

The tectonic evolution of the Central Mediterranean was partly controlled by the movement path of Africa relative to Eurasia, as shown in Figure 4.6: a SE-ward motion through the Jurassic and into the early Cretaceous, parallel to the Gibraltar transform fault, followed by an abrupt change in the mid-Cretaceous to northeast, then north-northeast into the Palaeogene, when it became northwards until the Miocene. Near the end of the Miocene it changed again to northwest, which is the present-day convergence direction.

However, the formation of the Alps and Apennines was controlled more by the movement paths of the independent terranes which, although guided in a general sense by the overall convergence of Africa and Eurasia, are constrained more by the geometry of the surrounding crustal blocks. Various reconstructions have been made of the geometry and movement history of these terranes, but the details, especially regarding the shapes of the terranes, must be regarded as speculative to some extent.

According to most interpretations, Gondwana separated from the southern margin of Eurasia during the Jurassic by the opening of the Piémont and Liguria ocean basins. These were bounded to the southwest by a transform fault running through Gibraltar, which connected these movements to the opening of the Central Atlantic. It is generally recognised that five separate independent terranes were involved in the further orogenic development of this region; these are coloured pink in Figure 4.6 (see previous chapter): Adria, Alcapa, Tisza, Alkapeca and the Briançonnais Terrane.

Figure 4.6 is based on a reconstruction of the possible sequence of events through the Cretaceous and Cenozoic. By the early Cretaceous, a small western branch of the Ligurian basin had opened east of what was then the eastern margin of Iberia, isolating the small terrane known as Alkapeca. Alcapa and Adria were still attached to Africa, but Tisza had separated from Alcapa in the north (Fig. 4.6A). During the mid-Cretaceous, the most southerly and largest terrane, Adria, consisting of the piece of continental crust now underlying eastern Italy and the Adriatic Sea, became separated from Africa by the Ionian Sea basin, a branch of the Neo-Tethys Ocean, and moved north to collide with Alcapa in what is known as the Eo-Alpine orogeny (Fig. 4.6B). Meanwhile, the northernmost terrane, Tisza, moved eastwards towards what became the Carpathians. Also during this period, the Briançonnais Terrane became separated from southeastern Europe by the small Valais Ocean basin. The Briançonnais was now flanked on its northwestern side by the Valais Ocean and on its southeastern by the Piémont–Liguria Ocean.

By the end of the Cretaceous, prompted by further expansion of the Ionian Sea, Adria–Alcapa had collided with the southern European margin in the sector that subsequently became the Eastern Alps (Fig. 4.6C). To the west, Alkapeca and the small Briançonnais terrane had collided with Iberia and southeast France respectively due to the closure of the two small ocean basins on their western flanks.

The critical stages in the tectonic evolution of the Alps occurred during the Oligocene to early Miocene Epochs. The change in convergence direction between Africa and Eurasia to a more northwards direction marked the early stages of the main collision event of the Western Alps (Fig. 4.6D) and was accompanied by extension in the Western Mediterranean with the opening of the Valencia and Balearic ocean basins (Fig. 5.3). This resulted in the south-eastwards translation of the Balearic Islands and the anti-clockwise rotation of Corsica and Sardinia into their present positions. These movements were accompanied by the subduction of the Ligurian ocean basin beneath the new eastern margin of Iberia, leading to the collision of Alkapeca (now welded to Iberia) with Adria to form the Apennines. At the same time, the Ionian Basin was consumed beneath the southern margin of Adria. During the later Neogene, a further phase of extension opened up the Tyrrhenian Sea basin between Corsica–Sardinia and the Italian Peninsula.

The formation of these ocean basins has been attributed to the process of back-arc extension on the upper plate of the Ionian Sea subduction zone, and is linked to the general eastwards retreat of the subduction zone due to trench roll-back (see Fig. 3.12). This has resulted in the subduction zone moving a distance of c.775km from its original position along the east side of Iberia in the Oligocene (where it resulted in the collapse of the Alborán Basin referred to in the previous chapter) to its present location in the Calabrian Arc (Fig. 5.3). The present-day position has been determined from the loci of active earthquakes and shows a near-vertical slab between 100 and 400km depth. The subduction zone is now very much shorter and more arcuate in shape, and its former continuation along the western side of the Adriatic is now a collision zone.

Tectonic structure

The broad geological architecture of the Western Alps has been known for well over a century. By the time

that Argand produced his great work in 1916 (e.g. see Fig. 5.2), much of the stratigraphic detail was already well known and the ideas of a European foreland and an 'African' hinterland that had converged to produce the visible complex structure were generally agreed, although many in the geological community were still opposed to continental drift. It is this western sector of the chain that is the most complex, and it is here that all the main tectonic zones are visible on a NW–SE traverse across it.

The main tectonic units of the Alpine chain, from north to south, are: the European Foreland, the Helvetic–Dauphinois (or external) Zone, the Pennine (or internal) Zone, the Austro-Alpine Nappes, the Southern Alps and the Adriatic Foreland (Fig. 5.4). At its widest, in the French-Swiss sector, the Alpine chain includes three additional zones: the Jura, the Foredeep (or Molasse) Basin and the Pre-Alps. Here the chain is nearly 300km across, although over most of its length is only about 200km across.

The European Foreland

The foreland consists of Palaeozoic basement belonging to the Eurasian Plate, which had previously experienced the Variscan orogeny, overlain by a Mesozoic sedimentary cover consisting of shallow-marine sediments dominated by shales and marls with prominent carbonate beds, consisting of both limestone and dolomite.

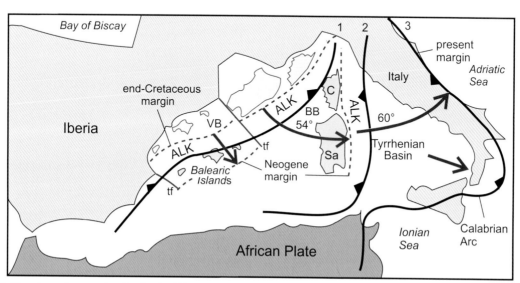

Figure 5.3 Opening of ocean basins in the Western Mediterranean. During the Neogene the subduction zone migrated eastwards due to trench roll-back. Note the former positions of the Alkapeca Terrane (ALK) on the upper plate of the subduction zone. Former position of islands in pale green; present positions in dark green. Note that Corsica and Sardinia have been rotated through 54° and the Italian Peninsula by a further 60°. BB, Balearic Basin; VB, Valencia Basin; C, Corsica; Sa, Sardinia; tf, transform fault. After Zeck, 1999.

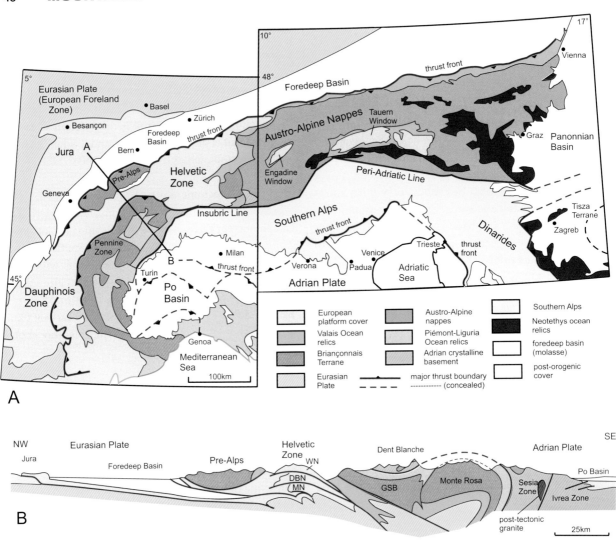

Figure 5.4 Main tectonic units of the Alps. **A**. Map. **B**. Diagrammatic section taken along line AB to illustrate the main structures. DBN, Dent Blanche Nappe; GSB, Grand St. Bernard Nappe; MN, Morcles Nappe; WN, Wildhorn Nappe. **C**. Schematic section showing an interpretation of the deep structure. A, C, after Schmid *et al.*, 2004; B, after Handy *et al.*, 2010.

The Jura, the Foredeep Basin and the Pre-Alps

The Mesozoic platform sediments are involved in a foreland fold-thrust belt, the outermost part of which forms the Jura Mountains. This prominent chain is only about 300km long and is separated from the main part of the fold-thrust belt by a foredeep basin containing non-marine clastic sediments (molasse) derived from the rising mountain chain. In the Swiss sector of the belt

there is a large outlier known as the Pre-Alps, consisting of folded and thrust rocks belonging to the Pennine Zone, which has become isolated from the main outcrop of that zone.

The Jura structure is the classic example of a so-called 'thin-skinned' fold-thrust belt, where a relatively thin sedimentary cover has been folded and moved along the basement (Fig. 5.5A). Albert Heim (1921) described

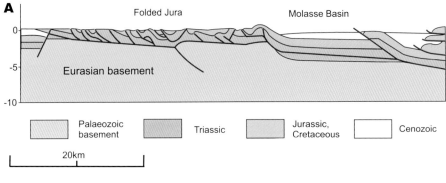

A

0
-5
-10

Folded Jura   Molasse Basin

Eurasian basement

Palaeozoic basement   Triassic   Jurassic, Cretaceous   Cenozoic

20km

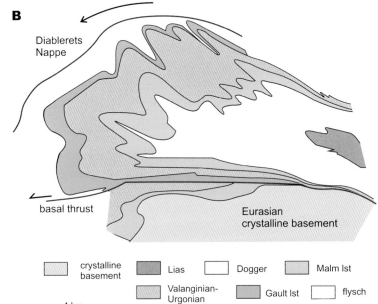

B

Diablerets Nappe

basal thrust   Eurasian crystalline basement

crystalline basement   Lias   Dogger   Malm lst

Valanginian-Urgonian   Gault lst   flysch

4 km

Figure 5.5 A. NW–SE cross-section of the Jura and Molasse Basin. Note that the anticlines correspond to topographic ridges and the synclines to valleys. The thrusts are mostly directed westwards towards the foreland and the basal thrust follows the top of the basement for much of its length within a weak Triassic evaporite. **B**. The Morcles Nappe: cross-section showing the typical ductile folding of the Dauphinois nappes as shown by the variation in thickness of the layers. The Diableret Nappe above has been arched over the Morcles nappe. A, after Pfiffner, 2014; B, after Dietrich & Song, 1984.

this as being like a table-cloth that has been pushed from one side and rumpled into ridges and valleys. The present topography reflects this structure in that the ridges are anticlines, often box-shaped, and the valleys are synclines – a topography that is well expressed in the hilly country between Bourg en Bresse and Geneva.

This type of structure requires the presence of a weak layer that allows the rocks above it to glide along it. Such a layer is termed a décollement (or detachment) surface, which in the case of the Jura, is provided by a Triassic evaporite bed consisting mostly of gypsum. The Mesozoic cover has been moved up to 7km along the top of the basement, which has remained almost undeformed. The basal thrust plane passes beneath the flat-lying sediments of the Molasse Basin to link with the thrusts in the interior of the belt, indicating that the compressive stress responsible for the Jura structure has been transferred from the more highly deformed nappes of the Dauphinois Zone.

The Foredeep Basin forms the low ground of the Swiss Plain, which extends from south of Geneva to Bern and thence eastwards along the front of the Austrian Alps. Here the concept of the 'molasse trough' was established, as a basin collecting mostly non-marine sediments derived from the mountain chain. These sediments range from Oligocene to earliest Pliocene in age and are dominated by coarse clastic deposits – conglomerates and sandstones – but also include layers of marine sandstones and shales. Although the term 'molasse' is generally used to denote a continental deposit, the environment of deposition is one where occasional flooding by a shallow sea brings marine deposits into the sequence. The thickness varies widely, up to over 4km towards the mountain front. The inner parts of the zone have been deformed by thrusting and folding, as shown in Figure 5.5A.

The Pre-Alps occupy the mountainous area about 110km long and 30km across, east and northeast of

Geneva. It is entirely surrounded by a thrust that is folded into a syncline and is detached from its continuation to the east in the Pennine Zone, as shown in the cross-section (Fig. 5.4B); it is composed of rocks identical to those of the Pennine Zone. Such a structure is known as a thrust outlier or klippe. The Pre-Alps are represented in Argand's cross-section as the outermost part of the East Alpine nappe system, which overlies the Pennine Zone as a continuous sheet, as in Figure 5.2. A more modern interpretation of the structure is that the pre-Alps forms a block that has become detached from the main Pennine nappes and slid onto the foreland under gravity due to the thickening and uplift of the underlying nappe pile.

The Helvetic–Dauphinois Zone

This zone corresponds to the main part of the foreland fold-thrust belt; in the outer 'autochthonous' part, the sedimentary cover is still attached to its basement but the inner part is said to be 'par-autochthonous' in the sense that the sedimentary sequence is still recognisably part of the foreland, although now completely detached from its basement. It consists of a set of complex fold nappes underlain by ductile thrusts, or shear zones (Fig. 5.5B); the cores of the nappes are composed of European Palaeozoic basement, much of which is represented by crystalline

massifs dominated by gneisses and granites such as the Mont Blanc massif (Fig. 5.6). The sedimentary cover of these massifs consists of a Mesozoic sequence including prominent carbonate beds, which can be correlated with those of the foreland, and which originated towards the margin of the European continental platform. Compared with the Jura, the Trias is much reduced in thickness, but the Jurassic sequence is much thicker, and a massive 300m-thick white Cretaceous limestone is prominent in the topography, enabling some of the complex fold structures to be easily traced out.

In the southeastern part of the zone, there are up to 2.5km of marine flysch sediments of Eocene age. This flysch sequence contains mass-flow deposits, known as turbidites, typical of the unstable environment of the continental slope. The significance of flysch deposits, first recognised in the Alps, in the historical investigation of orogenic belts was referred to in chapter 4.

The nappe structures are directed towards the foreland, as shown in Figure 5.4B, either westwards, north-westwards or northwards, depending on their position on the arc. The three main nappes in the Swiss sector are the Morcles, Dent Blanche and Wildhorn nappes: these form many of the high mountains of the Swiss Alps.

Figure 5.6 The Mont Blanc massif. These jagged peaks are carved from the Palaeozoic crystalline basement of the Helvetic–Dauphinois Zone. © Shutterstock, by Leonard Zhukovsky.

The outcrop width of the zone is greatest in the southwestern, French, sector, where it is known as the Dauphinois Zone. Here, southwest of Briançon, it reaches over 120km in width, but in the Austrian sector, the zone is almost entirely covered by the Austro-Alpine nappes except for a narrow strip along the thrust front and in two tectonic 'windows' where the cover has been stripped off.

The Pennine Zone

This is the internal, or 'allochthonous' zone, containing the metamorphic core of the mountain belt. The outcrop of the Pennine Zone is widest in the French-Swiss sector, where it occupies the high mountain belt straddling the French–Italian border from south of Briançon to Monte Rosa at the Swiss border, near Zermatt. It consists of three sub-zones: the Briançonnais Terrane, together with underlying and overlying zones of nappes containing oceanic deposits (Fig. 5.4B). The Briançonnais Terrane is only recognised in the Western Alps and must wedge out west of the Engadine Window in the Eastern Alps, where it is absent, although the other Pennine units continue eastwards. The frontal thrust of the Pennine Zone has travelled at least 40km over the External Zone in the Western Alps.

The Briançonnais Terrane consists of a Mesozoic sedimentary sequence underlain by Palaeozoic crystalline basement, which forms the core of several large nappes, including the Grand St Bernard and Monte Rosa Nappes (Fig. 5.7). The Mesozoic sequence differs significantly from that in the adjacent Helvetic–Dauphinois Zone. In the Vanoise region of eastern France, thick Permo-Carboniferous sediments, including coals, are overlain by a 300m-thick Triassic quartzite, succeeded in turn by up to 500m of carbonate. These are overlain by a much-reduced marine Jurassic–Cretaceous sequence containing evidence of several emergent episodes. The Eocene is represented by black pelagic shales followed by flysch.

The Briançonnais Terrane is underlain by a narrow zone of marine clastic deposits forming the lower Pennine nappes and overlain by the upper Pennine nappes. The latter contain a thick monotonous succession of mica-schists, known as the Schistes Lustrées, of Jurassic to Cretaceous age, which originated as calcareous shales and contain bands of basic volcanics and ophiolites.

These marked differences in stratigraphic history are explained by the sequence of events illustrated in Figure 4.6: the Briançonnais Terrane, having originated on a depressed European margin during the Triassic, broke

Figure 5.7 Monte Rosa. This peak, above Zermatt, is composed of Palaeozoic basement in the core of one of the Pennine nappes. © Shutterstock, by mountainpix.

away in the early Cretaceous, while the marine clastic sediments and ophiolites were formed in oceanic basins on each side of the terrane. The ophiolites and ocean-floor sediments of the lower Pennine nappes are thus thought to represent the remains of the Valais Ocean, and those of the upper Pennine Nappes to belong to the Piémont Ocean, which originally separated the Briançonnais Terrane from the Alcapa Terrane (see Fig. 4.6B).

The complex overfolds in the Pennine Zone take the form of a fan, directed towards the foreland in the northwest, like those of the underlying zones, but directed towards the southeast at the opposite side of the zone (Figs 5.4B, 5.8A). The nappes generally rest on a décollement surface of Triassic gypsum.

The rocks of the Pennine Zone have been affected by metamorphism of late Cretaceous or Palaeogene age. The grade of metamorphism varies from greenschist facies in the outer parts of the zone to blueschist and eclogite in the inner. The presence of blueschists and eclogites indicates that these parts of the zone were subjected to high pressures and low temperatures, which is interpreted as evidence that they must have been subducted to a considerable depth before being exhumed and thrust onto the European Plate. The low-pressure metamorphic

mineral assemblages were subsequently overprinted by higher-temperature, lower-pressure assemblages as the rocks were uplifted.

The Austro-Alpine Nappes

This zone consists of material that has long been interpreted as derived from the southern foreland, historically thought of as part of Africa but now assigned to the separate Alcapa Terrane. In the Eastern Alps, the Austro-Alpine Nappes form most of the outcrop area; the lower Pennine and Helvetic nappes only appear in the Engadine and Tauern Windows and in a narrow strip along the Alpine front. The Palaeozoic basement rocks of these nappes are overlain by a thick Permian to Triassic sedimentary cover dominated by massive shallow-marine limestones, which form the mountain chains known as the Northern Calcareous Alps. The oceanic material underlying these nappes thus represents the suture between Alcapa and the Eurasian plate.

In the Western Alps, tectonic units from the southern foreland only appear as high-level nappes such as the Dent Blanche Nappe, which is a klippe made of crystalline Adriatic basement overlying the Upper Pennine Nappes (Figs 5.4B, 5.8A). The Dent Blanche Nappe is considered

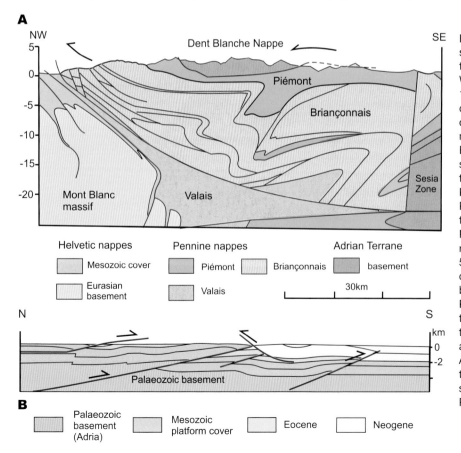

Figure 5.8 A. Pennine nappes: simplified cross-section through the Pennine Nappe Zone in Western Switzerland. Note: 1) the Lower Pennine nappe consisting of Valais Ocean crust is thrust over the Helvetic nappes. 2) The nappes of the Briançonnais Zone consist of a series of thrust slices overthrust towards the NW. 3) The Upper Pennine nappes consisting of Piémont Ocean crust overlie the Briançonnais Zone. 4) The Pennine nappes have been refolded and directed SE-wards. 5) The Dent Blanche Nappe consisting of Adrian crust has been thrust over the top of the Pennine Zone to form a klippe on the Dent Blanche. **B**. Structure of the Southern Alps: N–S section across the southern part of the Alto Adige north of Venice. Note the gentle open folding and south-directed thrusts. After Pfiffner, 2014

to have rooted in the steeply-dipping Sesia Zone, which forms the northern margin of the Adria Terrane.

The Adriatic Foreland and the Southern Alps

The foreland on the south (or east) side of the Alpine Chain consists of Palaeozoic crystalline basement belonging to the Adria Terrane. The Palaeozoic basement has long been regarded as being of 'African' origin; that is, it records a similar geological history to the adjacent parts of northern Gondwana. As described earlier, Adria was one of several such terranes that had broken away from Pangaea in the Jurassic or early Cretaceous (see Fig. 4.6B). The Adrian basement forms a small outcrop northeast of Turin, where it is known as the Ivrea Zone, but elsewhere is obscured by a Mesozoic to Cenozoic platform cover sequence that has been deformed into a typical foreland fold-thrust belt (Fig. 5.8B). This zone of southerly-directed folded and thrust rocks is known as the Southern Alps, which include the famous Dolomites mountain range (Fig. 5.9), formed from Mesozoic platform carbonates, and well-known to rock climbers. This fold belt is absent near Turin but broadens to form a 90km-wide belt north of Verona. The structure of the belt is rather subdued compared to that of the Western and Central Alps and the strata are often quite flat-lying.

The northern boundary of the Adrian Terrane is represented now by steep faults, the Insubric Line in the west and the Peri-Adriatic Line in the east. These 'lines' are late-orogenic strike-slip faults that have displaced the original inter-plate suture. North of these faults, the suture is folded at the base of the Austro-Alpine Nappes, but its position at depth on the southern side has been subject to various interpretations: Figure 5.4C, which is based on geophysical surveying, shows the inter-plate boundary dipping southwards, but distorted by southerly-directed back-folding. However, more easterly cross-sections have been interpreted differently.

The Po Basin

The folded Mesozoic cover of the Southern Alps is overlain by the Po Basin, which is a foredeep basin consisting of several kilometres of Cenozoic sediments derived from the rising Alps. This basin received sediments from all three of the fold belts that surround the basin: the Apennines, Alps and Dinarides. The centre of the basin contains marine sediments of Eocene to Miocene age, but towards the Alpine margins, the marine strata give way to up to 3.6km of non-marine molasse, dominated by late Miocene to Oligocene conglomerates.

Tectonic summary

The structures of the outermost zones, the folded Jura and the Southern Alps, exhibit typical fold-thrust geometry

Figure 5.9 The Dolomites. The Odle Mountains from Santa Maddalena in Val di Funes. These massive carbonate beds are a prominent feature of the Southern Alps. © Shutterstock, by Ttstudio.

(Figs 5.5A, 5.8B), with outward-directed thrusts and overfolds linked to a basal thrust along a weak stratigraphic horizon. The nappes of the Helvetic zone, having originated at greater depths, are more ductile, but still have an overall, foreland-directed, overthrust sense of movement (Fig. 5.5B). The more complex nature of the folding in the Pennine Zone, illustrated in Figure 5.8A, has been attributed to SSE-directed back-thrusting. This has had the effect of moving the Pennine zone upwards and backwards towards the Adrian Terrane, thus raising the metamorphosed interior of the orogenic belt to a higher level and isolating the Dent Blanche nappe (see Fig. 5.4B). This back-thrusting may have been caused by continued convergence between the opposing continental plates at a time when further overthrusting was prevented by the increased thickness of the belt.

Detailed analysis of the ductile structures of the Pennine zone indicate that the movement direction changed through time in an anti-clockwise sense, suggesting a link with the anti-clockwise change in the plate convergence direction noted previously. A similar pattern has been observed in the outer nappes of the Helvetic–Dauphinois Zone, with the outer, younger, thrusts showing a more northwesterly movement direction. The change in convergence direction was a response to the anti-clockwise rotation of Adria brought about by the expansion of first, the Balearic Basin, and then the Tyrrhenian Basin. Gravity gliding is thought to be responsible for the isolated position of the Pre-Alps and for many of the nappes in the Helvetic–Dauphinois Zone.

Accurate measurement of the total amount of shortening across the Alpine belt is impossible because of the complexity of the structures, especially in the interior zones, but it is likely that the shortening has been in excess of 250km across the belt. This compares with a crustal thickening to over 50km.

Tectonic history

Crustal deformation commenced during the mid-Cretaceous, as a result of the convergence between the Adrian and Alcapan terranes, resulting in the overthrusting of Adria onto the margin of the Alcapa Terrane – an event known as the Eo-Alpine Orogeny. This event was confined to the Austro-Alpine nappes. Subduction of the Piémont Ocean crust along the northern margin of Alcapa led eventually to the combined Adria–Alcapa microplate colliding with the Eurasian margin in the late Cretaceous. Alcapa was overthrust onto Eurasia to form the Austro-Alpine Nappes.

Further west, during the Palaeogene, as a result of the subduction of the Valais Ocean, the Briançonnais Terrane collided with the Eurasian margin. The main Alpine orogeny commenced during the Eocene when the Ligurian Ocean crust separating Adria from the Briançonnais Terrane, now welded to the European margin, was subducted beneath Adria. This resulted in the overthrusting of Adria onto the Briançonnais Terrane, and caused the already deformed Austro-Alpine Nappes to be emplaced over the European foreland.

The complex shape of both plate margins, together with the change in the convergence direction noted above, resulted in a significant regional variation in the orientation of the structures. The main Alpine deformation commenced in the Pennine Zone and progressed outwards to the Helvetic–Dauphinois Zone and onto the Foreland. Crustal shortening and uplift continued through the Oligocene, reaching its climax in the late Oligocene, about 25Ma ago, and continuing into the Miocene Epoch with the late back-thrusting phase. However, some convergent movement and uplift still continues today.

The Apennines

The Apennine Mountains consist of a series of mountain ranges extending along the spine of the Italian Peninsula from near Genoa in the northwest to the extreme southwest end of the 'toe' of Italy, in Calabria, a distance of 1200km (Fig. 5.10). The eastern slopes of the Apennines are inclined steeply down towards the Adriatic Sea while the western slopes are generally gentler and host many fertile valleys. The mountain system as a whole is relatively narrow compared with the Alps, and never more than about 250km across. It contains thirteen peaks over 2000m and is divided naturally into three sectors. The northern sector, aligned NW–SE, runs from near Genoa to east of Florence; the central sector, northeast of Rome, runs parallel to the Adriatic coast, and contains the highest peaks, including the highest, Corno Grande (2912m) in the Gran Sasso d'Italia Massif (Fig. 5.11). The northern and central sectors together form what is known geologically as the Northern Apennine Arc, ending north of the Bay of Naples. The southern sector, or Southern Apennine Arc, forms an arc convex towards the Adriatic, and curves round into the end of the Calabrian peninsula to continue across northern Sicily; these mountains are lower than those of the central sector, none being higher than 2000m.

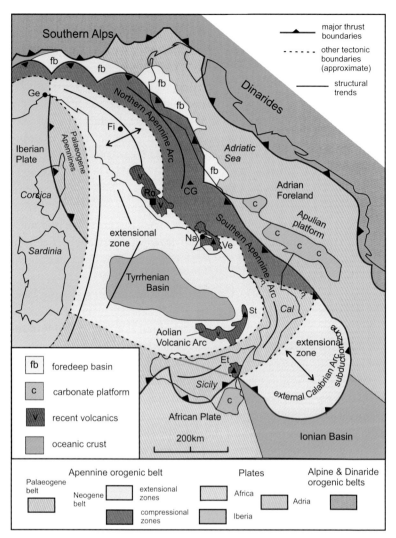

Figure 5.10 The Apennines. Simplified tectonic map of the Apennine orogenic belt showing its relationship to the Alpine belt and the adjoining Iberian, African and Adrian plates. Note: 1) the older Palaeogene belt in the west and south has been separated from the younger Neogene belt in the east by a wide extensional region, much of which now forms a marine basin; 2) the active subduction zone in the southeast producing the Aeolian Volcanic Arc has formed a wide extensional zone on its upper plate. Cal, Calabria; Fi, Firenze (Florence); Ge, Genoa; Na, Naples; Ro, Rome. CG, Corno Grande; Et, Etna; St, Stromboli; Ve, Vesuvius. After Patacca & Scandone, 2007.

Tectonic setting

The Apennines appear to continue from the southwest end of the Alps (Fig. 5.1). However, although at first sight connected, their origin is quite different from that of the Alps. While the Alpine orogenic belt was forming along the northern boundary of Adria, the western side of Adria, where the future Apennines would form, was facing the Ligurian Ocean (see Fig. 4.6C).

The upper plate of the Apennine collisional system consists of the Alkapeca Terrane, formerly part of the Iberian Plate (see Fig. 4.6A). As we saw earlier, this terrane was separated from the Iberian Plate by the western branch of the Ligurian Ocean in the Cretaceous. This ocean began to close in the late Cretaceous by eastwards subduction beneath Alkapeca; closure was completed during the mid-Palaeogene, Alkapeca having been thrust westwards onto Corsica, creating the collisional belt of the Western, or Palaeogene Apennines to form a branch

of the Alpine orogenic belt that extended around the eastern rim of Iberia (see Fig. 4.6C).

Deformation continued on the eastern side of this belt as subduction of the eastern Ligurian Ocean commenced along the eastern side of the combined Iberian–Alkapecan Plate. Adria and Iberia converged during the late Palaeogene, and during the early Miocene, around 20Ma ago, Alkapeca became sandwiched between Corsica–Sardinia, at the eastern edge of Iberia, and Adria. Adria thus formed the lower plate of a subduction zone that extended along the eastern side of the Italian Peninsula, and when the two pieces of continental crust collided, Alkapeca was backthrust onto Adria. (see Fig. 4.6D). Deformation continued until the present day with further convergence between Alkapeca and the Adrian Foreland.

The south-eastwards progression of the subduction zone is tracked by the changes in the ages of the volcanism from northwest to southeast. The volcanism in Sardinia and in the islands off the northwest coast of Italy is of late Miocene age, whereas that on the Italian mainland, including Vesuvius, is of Pliocene age. The youngest vulcanicity is in the currently active Aeolian Arc, north of Sicily. Seismic information from earthquakes associated with the subduction process can be used to trace the position of the present-day subducting slab, which outcrops in a wide arc southeast of Sicily and Calabria and reaches a depth of 450km beneath the Tyrrhenian Sea (Fig. 5.10).

Tectonic structure

The western part of the Apennine system (the Palaeogene Apennines) consists of Alkapecan crust that had been thrust over the eastern part of the Iberian Plate during the Palaeogene and is now exposed in eastern Corsica, Sicily, Calabria and the western coastal sector of the Northern Apennines (Figs 5.10, 5.12). Much of this sector is composed of Alkapecan Palaeozoic basement. The piece of Alkapecan crust outcropping in eastern Corsica has experienced high-pressure metamorphism of late Cretaceous to early Palaeogene age overprinted by low-temperature metamorphism during Neogene extension.

The eastern part of the Apennine system is a foreland fold-thrust belt, consisting of thrust slices of Adrian platform cover overthrust eastwards onto the Adrian Plate. In the northern and central Apennines, this compressional belt is separated from the western side of the orogen by a wide extensional zone, much of which is submerged beneath the Tyrrhenian Sea. Figure 5.10 shows the eastern boundary of the Apennines, now a thrust front, following the east side of the Italian Peninsula to Apulia. Here it joins the still-active part of the subduction zone, which forms the semi-circular Calabrian Arc along which Neo-Tethys Ocean crust of the Ionian Basin descends beneath Italy. This subduction zone has given rise to a volcanic arc on its upper plate that includes the three famous Italian volcanoes: Vesuvius, Etna, and Stromboli, the latter belonging to the Aeolian Islands Volcanic Arc, which mostly consists of extinct volcanoes. The Apennine thrust front reappears in Sicily, where it forms the northern margin of the African Plate, and crosses to Africa where it joins the edge of the Atlas Orogenic Belt described in the previous chapter.

In both Alkapeca and Adria, the Palaeozoic basement is overlain by a sedimentary cover of Mesozoic age, which had been deposited along the margins of the Tethys Ocean. Much of this sedimentary cover consists of a carbonate platform of Triassic to Palaeogene age that is now exposed on the Apulian Platform, the southeastern coastal plain that extends to the end of the 'heel' of the Italian peninsula. This cover was involved in the fold-thrust belt on the Adriatic side of the orogen together with the overlying Neogene sedimentary cover deposited on the active margin. The lower units of the thrust stack consist of Adrian Platform cover, dominated by carbonates, while the upper units are derived from deeper-water basins originating further to the west. These include flysch deposits from the ocean trench, which,

Figure 5.11 The Pizzo d'intermesoli and surrounding mountains of the Gran Sasso Massif, Central Apennines. Note the uniformly inclined bedding of the Mesozoic sediments. © Shutterstock, by Eder.

together with the deformed platform strata, make up the accretionary prism. A series of foredeep basins surround the northern and eastern sides of the Northern Apennine Arc. These contain a thick succession of Pliocene to Pleistocene clastic sediments derived from the rising mountain chain.

Tectonic history

The Apennine belt differs significantly from the other parts of the Alpine–Himalayan Belt so far described, in that such a large proportion of the belt, as originally formed, is now dominated by extensional structures. While the eastern side of the Apennine chain is a compressional fold-thrust belt, the western side is dominated by extensional structures in the form of a system of fault blocks (Fig. 5.12). The extensional zone has split the older Palaeogene belt into two separate parts, as shown in Figure 5.10: one in the northwest, in Eastern Corsica, and the second in the south, extending from Calabria through Sicily to Tunisia. This system reflects the back-arc extensional regime on the upper plate of the Adrian subduction zone, which, in its central part, created the oceanic basin of the Tyrrhenian Sea during the late Miocene Epoch. There is no clear boundary between the compressional and extensional zones, and most of the original thrusts in the western part of the orogen have been re-activated in extension, as shown in Figure 5.12. It is thought that the two processes may have co-existed to some extent, so that while the lower parts of the orogen were still subject to convergence, the upper parts may have been dominated by extension.

The history of subduction in the Italian Peninsula is part of the regional process illustrated in Figure 5.3, involving the gradual eastwards retreat of the subduction zone from its original Palaeogene position around the eastern and southern sides of the combined Iberian–Alkapecan Plate to its present position in the Calabrian Arc – a distance of around 1200km. This was accommodated by the creation of the large extensional basins of the Western Mediterranean, such as the Balearic and Tyrrhenian Seas. As a consequence of this eastwards movement, Corsica–Sardinia experienced a counter-clockwise rotation of 54° between the late Eocene and the mid-Miocene, and the Italian Peninsula by a further 60° from the late Miocene to the present day.

The cause of back-arc extension is thought to lie in the process of 'trench roll-back', described in chapter 3 (see Fig. 3.12), where the position of the down-bend of the subducting slab (which determines the position of the trench) retreats backwards due to the downward pull of the cold, heavy, slab. It has also been suggested that a piece or pieces of lithospheric mantle may have become detached from the upper plate and sunk, leaving the crust above warmer, weaker, and more subject to extension. The precise way in which these processes may have worked in the Mediterranean context would have been governed by processes in the mantle that are not well understood, but must depend to some extent on how the descending slab was constrained by the geometry of the surrounding continental lithosphere – Eurasia to the north and Africa to the south – which were converging during this period and exerting a lateral pressure on the subducting slab.

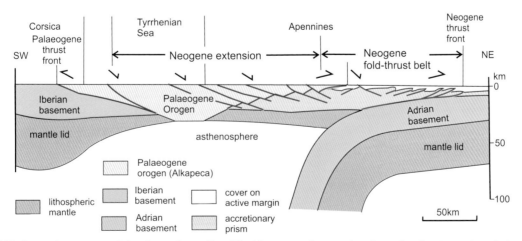

Figure 5.12 Crustal structure of the Apennines. Simplified interpretative section from Corsica to eastern Italy showing collisional structures formed by the thrusting of Alkapeca westwards over Iberian basement and eastwards over Adria. These are cut by extensional structures associated with the opening of the Tyrrhenian Sea. After Carminati & Doglioni, 2012.

6

The Carpathians, the Balkans and Turkey

The Alpine orogenic system east of the Alps divides into two branches: the Carpathians, which describe a great arc around the Pannonian Basin, and the Dinarides, which continue down the eastern side of the Adriatic Sea through the Balkans and into Greece, where it is known as the Hellenides (Fig. 6.1; and see Fig. 5.1). The southwestern boundary of the Hellenide sector of the orogenic belt is defined by a subduction zone, along which Neo-Tethys Ocean crust of the Eastern Mediterranean is being consumed, while its northern boundary is the frontal thrust along the southern margin of the Moesian Platform. The total width of the Alpine belt here is around 1600km, but this gives a misleading impression of the amount of deformation across the belt, since a large part of it is underlain by extended crust, now largely submerged beneath the Aegean Sea, and by the Rhodope Massif, which suffered comparatively little Alpine deformation. The belt here is divided into two parts by the Vardar Suture, which marks the site of the former Neo-Tethys Ocean and separates African-derived terranes to the west and south from the Eurasian-derived Rhodope Massif.

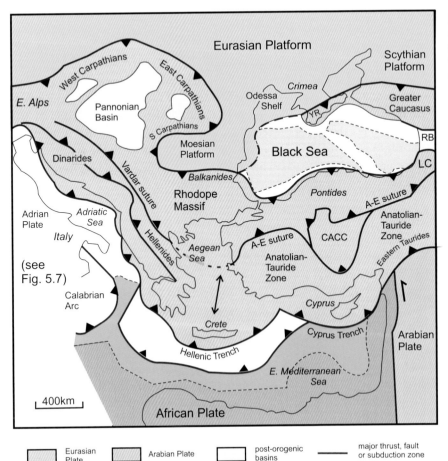

Figure 6.1 Alpine orogenic belts of the Eastern Mediterranean and Black Sea Region. CACC, Central Anatolian Crystalline Complex; A–E, Ankara–Erzincan; LC, Lesser Caucasus; RB, Rioni Basin; YR, Yayla Range. After Okay, 2000.

South of the Black Sea, the southern, and greater, part of the Alpine belt continues through Turkey, while the northern marginal zone is submerged beneath the northwestern part of the Black Sea, to re-emerge in the Crimean Peninsula and continue eastwards as the Greater Caucasus, which extends from the northern shores of the Black Sea towards the Caspian Sea.

In Asian Turkey, the Alpine belt is divided into the complex Pontide zone in the north and the Anatolian–Tauride zone in the south, which are separated by the Ankara–Erzincan Suture, the eastwards continuation of the Vardar Suture in Greece. The Pontide Zone has Eurasian affinities similar to Dacia and the Rhodope massif, whereas the Anatolian–Tauride Terrane was derived from the African–Arabian Plate in the early Cretaceous and collided with the Pontide Zone during the Palaeogene. A large central block, known as the Central Anatolian Crystalline Complex, also of Gondwanan origin, became sandwiched between the two. The southern margin of the belt here is defined by

an arcuate subduction zone situated south of Cyprus. Much of the Alpine deformation was concentrated along the contact zones between the main terranes, and in the Eastern Taurides, where the Arabian Plate converged with Laurasia. The suture zones are distinguished by extensive ophiolites. The Pontide Zone continues eastwards into the Lesser Caucasus, which is separated from the Greater Caucasus by post-orogenic sedimentary basins.

The Carpathians

The Carpathian mountain chain describes a wide semicircle around the Pannonian Basin and is divided into three branches: the Western, Eastern and Southern Carpathians (Figs 6.1, 6.2). With a total length of 1500km, it is nearly as long as the Alps, but is much less impressive in terms of height. The western end of the chain, in Slovakia, is separated from the eastern end of the Alps by the Danube valley at Bratislava, and

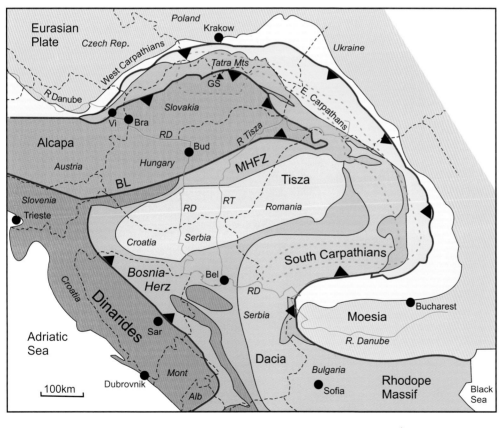

Figure 6.2 The Carpathian Region. Simplified map showing the main tectonic units. BL, Insubric–Balaton Line; MHFZ, Mid-Hungarian Fault Zone. Countries: Alb, Albania; Mont, Montenegro. Cities: Bel, Belgrade; Bra, Bratislava; Bud, Budapest; Sar, Sarajevo; Vi, Vienna. Rivers: RD, River Danube; RT, River Tisza. After Márton et al., 2007.

from there it extends eastwards to form the boundary between Slovakia and Poland, where the chain is known as the Tatra Mountains. Here the Western Carpathians are at their widest and contain the highest peaks, the highest being Gerlachovský štít, at 2655m (Fig. 6.3). The Eastern Carpathians describe a wide arc from south of Krakow in southeast Poland through southwestern Ukraine into Romania, then turn abruptly west into the Southern Carpathians, otherwise known as the Transylvanian Alps. Here the chain describes an arc that is convex inwards towards the Hungarian Plain and contains several summits over 2500m. The chain ends at the Serbian border where it is cut through again by the River Danube.

Tectonic structure

Each of the three branches of the Carpathians – the Western, Eastern and Southern – has the same basic structure: the outer foreland, the foredeep basin, the main Carpathian thrust belt, and the interior terranes of the Pannonian Basin (Fig. 6.2). The boundaries of the interior terranes are distinguished by zones of oceanic material, including ophiolites that mark the sites of the former ocean that separated the interior terranes from Eurasia.

The Outer Foreland

In the case of the Western and Eastern Carpathians, the foreland is part of the Eurasian Platform, and consists of Palaeozoic basement with a Mesozoic platform cover very similar to that of the Alps. The southern branch, however, is bounded by the Moesian Terrane, or Moesian Platform, which is also of Eurasian origin, but is separated from the main part of the Eurasian Plate by a narrow Triassic rift (not shown in Fig. 6.2).

The Foredeep Basin

The outermost zone of the Carpathian belt consists of a molasse-filled trough up to 100km wide, overlying the foreland margin.

The Carpathian Thrust Belt

This belt is up to 100km wide in the Eastern Carpathians but narrows to the west and south, where it is overlain by younger deposits, and is divided into outer and inner zones. The outer zone consists of a series of Palaeogene to early Neogene thin-skinned thrust nappes directed outwards towards the foreland, and containing a variety of rock types, including both sedimentary units from the Eurasian platform cover and flysch units derived from the ocean that originally separated the interior

Figure 6.3 Gerlachovsky štít: highest peak in the Tatra Range, Eastern Carpathians, northern Slovenia. © Shutterstock, by Jaroslav Moravcik.

terranes from the Eurasian margin. The inner zone consists of nappes of Gondwanan basement and Mesozoic platform cover belonging to the three interior terranes of the Pannonian Basin. There is widespread evidence of calc-alkaline (andesitic) vulcanicity that was active from the Miocene to the present day. The thrusting appears to have progressed outwards towards the Eurasian foreland up to Pliocene times.

The oceanic relics

These consist of a variety of rocks, including ophiolites, considered to have been derived from an accretionary complex. They are contained within thrust slices that tectonically overlie the Thrust Belt and mark the former position of a subduction zone dipping inwards towards the Pannonian Basin. The Eurasian Foreland was thus a passive margin and the oceanic zone, site of the former Meliata–Vardar Ocean, marks the suture between it and the terranes of the interior.

The Pannonian Basin

Three separate terranes – Alcapa, Tisza and Dacia – have been identified within the basement complex of the Pannonian Basin, but are largely concealed beneath the Neogene cover. These terranes are separated by suture zones featuring ophiolites and other oceanic relics, indicating that they would have moved independently. Alcapa and Tisza were discussed previously in the context of the Alps. 'Alcapa', named from Alps–Carpathians–Pannonia, forms the Austro-Alpine Nappes (see Fig. 5.4) and extends eastwards into the Western Carpathians, where it forms the basement of the northern sector of the Pannonian Basin. The Tisza Terrane, named after the River Tisza that flows through the centre of the basin, lies immediately south of the Alcapa Terrane and is separated from it by the Mid-Hungarian Fault Zone, which includes the important Balaton Line. This lineament continues westwards into the Alps to join the Peri-Adriatic Line, where it marks the boundary between the Austro-Alpine Nappes (derived from Alcapa) and the Adrian Plate (see Fig. 5.4). Dacia lies south of Tisza and wraps around the Moesian promontory, continuing south-eastwards as the Rhodope Massif.

The crust underlying the Pannonian Basin experienced significant east-west stretching, crustal thinning, and extensive vulcanicity during the Neogene, resulting in the formation of the depression in which 6–8km of molasse deposits have accumulated.

Plate-tectonic history

The origin of the Carpathians is attributed to the movement of four independent terranes, Adria, Alcapa, Tisza and Dacia, the last three of which have acted as indenters, driving in a northeasterly direction into the embayment created by the Eurasian Plate margin. Two of these terranes, Alcapa and Tisza, both of Gondwanan affinities, were discussed previously in the context of the origin of the Alps. These terranes were separated from the southern margin of Europe, while it was still attached to Gondwana, by the Piémont and Vardar oceans during the Jurassic, as shown in Figure 4.6A.

Dacia is believed to have been detached from further east along the European margin, also during the Jurassic. It forms part of an elongate strip of European basement that now curves through 180°, wrapping around the promontory of the Moesian Platform through the Balkans to the Rhodope Massif in northern Greece (Fig. 6.2).

During the Cretaceous, the northern and eastern margins of Alcapa, Tisza and Dacia were defined by subduction zones along which the Meliata–Vardar (Neo-Tethys) Ocean was consumed. This eventually resulted in Tisza being thrust over Dacia, which was in turn thrust over the European margin. Also during the Cretaceous, the large Adria Terrane moved north, collided with Alcapa, and then with the southern European margin in the late Cretaceous. The sequence of these movements is summarised in Figure 4.5.

During the Oligocene and Miocene, the subduction zone seems to have retreated eastwards due to slab roll-back into the Pannonian embayment, pulling the upper-plate Alcapa and Tisza terranes eastwards, and resulting in the Alcapa Terrane being thrust over Tisza. The position of the subduction zone was defined by an andesitic volcanic arc west of the Eastern Carpathians. The current position of the slab has been imaged by a concentration of deep-focus earthquakes in the southeastern corner of the orogenic belt, where the eastern and southern Carpathians meet.

The combined effect of the arrival of the Alcapa Terrane in the north, the eastwards push from the Adriatic Terrane in the west to form the Dinarides, and the trench roll-back in the early Neogene seems to have resulted in squeezing the other two terranes (Tisza–Dacia) into the U-shaped embayment in the European margin with the westward-projecting Moesian Platform on its southern side, and in the process has rotated and distorted them. Palaeomagnetic measurements on late Cretaceous and Palaeogene units indicate a large clockwise rotation of 120° in Tisza–Dacia from the late Palaeogene, which is consistent with the above interpretation, but also record an anti-clockwise rotation of 80° in Alcapa during the Miocene, suggesting that the clockwise rotation of Tisza–Dacia around the Moesian promontory caused a counter rotation in the adjoining Alcapa Terrane. The Alcapa–Tisza thrust boundary, originally oriented E–W to NW–SE and north-facing in the Oligocene, has thus rotated into its present WSW–ESE orientation, facing south, in the Miocene.

The Dinarides

The Dinarides, or Dinaric Alps, extend down the eastern side of the Adriatic Sea through the Balkans (Figs 6.1, 6.2) from Slovenia in the north to Albania in the south, crossing parts of Croatia, Bosnia-Herzegovina, Montenegro and Albania. The northern part of the belt runs in a NW–SE direction for about 620km, then turns southwards for a further 240km through Albania. The mountain belt then continues through Greece, where it is known as the Hellenides. The Dinarides thus form

a mountain barrier on the southwestern side of the Pannonian Basin, closing off the U-shaped loop formed by the Carpathians.

The Dinaride chain is relatively narrow in the northwest and southeast, but swells out in Bosnia to around 200km in width before narrowing again through Albania, and is never especially high, averaging between 1000 and 2000 metres in elevation, with only a few peaks over 2500m. The highest peak is Mount Prokletije, at 2694m. The range is dominated by carbonate rocks, and the Dalmatian region in southern Croatia is famous for its limestone scenery; it hosts the type locality of the karst landscape (Fig. 6.4), with its carbonate-solution features such as underground drainage, caves and sinkholes.

Tectonic structure

In tectonic terms, the Dinarides is an Alpine collisional belt forming the eastern margin of the Adrian Plate (Fig. 6.5A) and continues southwards into Greece as the Hellenides. The northwestern sector of the belt, which is approximately 300km across, consists of three main tectonic zones: the Adrian Platform, the Western Thrust Belt, and the Eastern Thrust Belt (Fig. 6.5B). The

Figure 6.4 Karst landscape of Dalmatia, Northern Dinarides. The Zrmanja River canyon, Croatia. Shutterstock©Tamisklao.

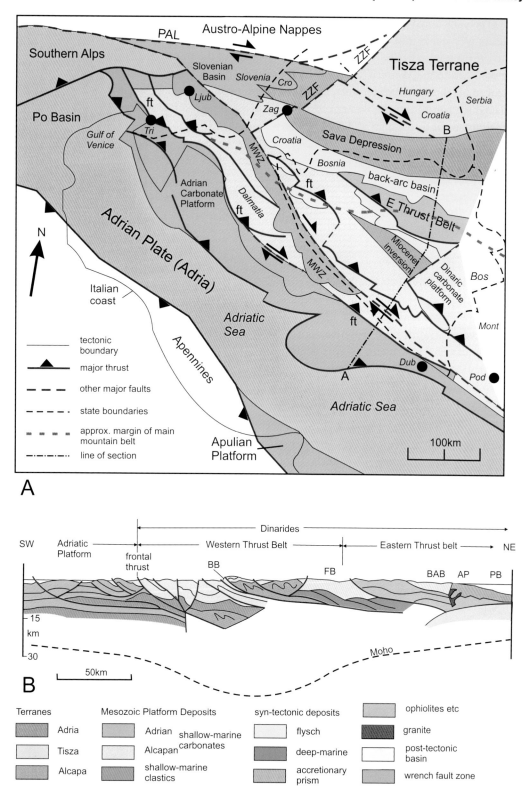

Figure 6.5 Major tectonic features of the Dinarides. **A**. Map: MWZ, Miocene wrench zone; PAL, Peri-Adriatic Lineament; ft, frontal thrust of Western Thrust Belt; ZZF, Zagreb-Zemplin Fault. A-B, line of cross-section. States: Bos, Bosnia; Cro, Croatia; Mont, Montenegro. Cities: Dub, Dubrovnic; Lub, Lubljana; Pod, Podgorica; Tri, Trieste; Zag, Zagreb. **B**. Cross-section: AP, accretionary prism; BAB, back-arc basin; BB, Budva Basin; FB, foreland basin; PB, Pannonian Basin. After Tari, 2002.

orogenic belt is actually much wider than the mountain range, extending westwards into the Adriatic Sea and eastwards into the Pannonian Basin. The mountain range is essentially confined to the Western Thrust Belt. The Dinarides incorporate three separate terranes, from west to east: Adria, which constitutes the western foreland; Alcapa, forming the main thrust belt; and Tisza, which is the eastern foreland to the belt but is mostly concealed beneath the sedimentary cover of the Pannonian basin.

The Eastern Thrust Belt is essentially a Mesozoic subduction belt culminating in late Cretaceous collision, while the Western Thrust Belt formed as a result of a later collision event during the Miocene.

The Adrian Platform

This zone is a foreland thrust belt involving the massive carbonate platform cover of the Adrian Plate, and includes the karst terrains referred to earlier. The thrusts are mainly directed westwards onto the Adriatic foreland, although limited back-thrusting is also evident. The zone occupies the coastal belt of Dalmatia in southwestern Croatia, including its numerous islands, and is partly submerged beneath the Adriatic Sea.

The Western Thrust Belt

Three subdivisions are recognised in this zone (Fig. 6.5B). The westernmost sub-zone consists of SW-directed folded and thrust units derived from the carbonate

platform of the Adrian Terrane, known as the Dinaric Platform. The frontal thrust has over-ridden the oceanic rocks of the Budva Basin, which originally separated the Dinaric Platform from Pelagonia (Fig. 6.6). The central unit contains a thin Mesozoic sedimentary cover overlying Palaeozoic basement, which has been overthrust onto the Dinaric Platform. The eastern unit is dominated by a flysch-filled foredeep basin of Cretaceous age overlying older deeper-marine sediments.

The Eastern Thrust Belt

This zone contains a set of westwards-directed thrust nappes consisting of a Jurassic to Palaeogene sequence, including an ophiolite-bearing unit containing oceanic crustal and upper-mantle rocks. In the southern sector of the zone, the ophiolites are overlain by late Cretaceous to Palaeogene flysch interpreted as a back-arc basin deposit. The northeastern part of the zone contains a unit known as the Sava Depression, interpreted as an accretionary prism resulting from the subduction of Vardar oceanic crust beneath the Dinaric Platform. The northern margin of the zone is overlain by Neogene molasse of the Pannonian Basin.

Tectonic history

The origin of the Dinarides lies in the relative movements of Adria, Alcapa and Tisza–Dacia. In the Triassic, Adria, Alcapa and Tisza, all parts of Gondwana, were

Figure 6.6 The Dinaride region in the early Triassic: reconstruction of the plate-tectonic elements. After Tari, 2002.

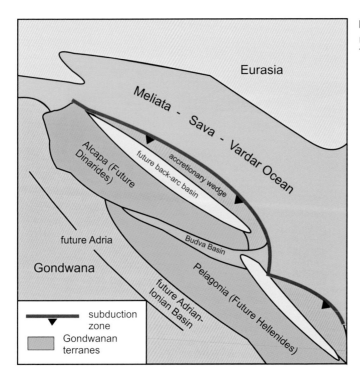

separated from southern Eurasia by the Neo-Tethys Ocean (see Fig. 4.6). By this time most of the original Tethys Ocean had been subducted and replaced by branches of the Neo-Tethys Ocean, variously called the Meliata, Sava or Vardar Ocean, depending on in which sector of the orogenic belt the remnants are now found. Southwest-directed subduction of the Vardar Ocean crust along the eastern border of Alcapa from Triassic times led to the development of extensional back-arc basins on the northeastern margin of the Adria–Alcapa platform, one of which, the Budva Basin, separated what became the Dinarides from Pelagonia in the future Hellenides (Fig. 6.6). These extensional basins gave rise in the Jurassic to the ophiolites of the Eastern Thrust Belt. Also during the Jurassic, as explained above in discussing the origin of the Carpathians, both the Tisza and Dacia Terranes had also become detached from Eurasia by the formation of narrow oceanic basins, and by the early Cretaceous, Tisza had collided with Dacia and the combined Tisza–Dacia block then formed part of the Eurasian foreland to the evolving Dinarides.

From the late Jurassic through the Cretaceous, convergence between Alcapa and Tisza–Dacia was accommodated by subduction along the Alcapan margin, accompanied by west-directed thrusting and the development of an accretionary prism dominated by flysch deposits. On the eastern (Tisza) side of the Dinaride orogen, east-directed thrusting affected the accretionary prism.

During the early- to mid-Cretaceous, Vardar Ocean crust was being subducted beneath all three Gondwanan terranes, which were separated by narrow marine basins. Convergence between Adria and Alcapa in the mid-Cretaceous produced the Eo-Alpine collision belt, part of which subsequently became the Austro-Alpine belt of the Eastern Alps. Final closure of the Vardar Ocean near the end of the Cretaceous led to the collision between the Alcapa and Tisza Terranes. The westwards migration of the thrust front led to the imbrication of the Alcapa Platform, the uplift of the thrust belt, and the development of a foredeep basin beyond the thrust front, which also migrated towards the foreland.

The Main Dinaride orogenic event took place in the late Oligocene to early Miocene in response to the opening of the Western Mediterranean Basins, which resulted in the eastwards translation and counter-clockwise rotation of Adria, as indicated in Figure 5.3. This culminated in the under-thrusting of Adria beneath the Dinarides, the closure of the Budva Basin, and the formation and subsequent uplift of the Western Thrust Belt.

The Neogene movements were also responsible for two sets of dextral wrench faults: a more prominent NW–SE set parallel to the main tectonic boundaries, and a lesser NE–SW set that includes the Zagreb–Zemplin fault defining the northwest boundary of the Tisza Terrane. Extension and subsidence of the Pannonian Basin allowed the Sava accretionary prism and much of the Tisza basement to be overlain by thick molasse deposits. Movements continued through the Neogene, and the belt is still seismically active.

The Hellenides and the Rhodope Massif

The continuation of the Alpine Orogenic Belt south-eastwards into Albania and Greece is known as the Hellenides (Fig. 6.1). The belt as a whole is up to 300km wide here, and contains the prominent Pindos Mountains (Figs 6.7, 6.8), which have been described as the 'spine' of Greece, running through the middle of the Greek Peninsula and continuing through the Peloponnese. The Pindos Range is only 50–60km across and is typically between 1000 and 2000m in height; the highest peak is Mt. Smolikas (2637m), near the Albanian border. The highest mountain in Greece, Mt. Olympus, at 2911m, is not in the Pindos Range, but north of it, on the western shores of the Aegean Sea.

The Rhodope Massif is a mountain region over 200km in width, lying north of the Hellenides, mostly in southern Bulgaria, but extending into Thrace in northern Greece. It contains two separate mountain ranges: the Balkan Mountains, (or Stara Planina) in the north and the Rhodope Range in the south. The Balkan Mountains are the continuation of the south Carpathian fold belt discussed earlier, and separate the main part of the Rhodope Massif from the Moesian Platform in the north. This mountain range is only about 40km wide, and is not especially high; the Rhodope Range is higher and much wider, forming the greater part of the Massif. The highest peak in the Rhodopes is Musala, at 2925m.

Tectonic structure of the Hellenide–Rhodope Belt

The continuation of the Alpine Orogenic Belt into Greece is characterised by significant structural changes. In contrast to the Dinarides, which is a collisional belt, the Hellenides is an active continental margin with a subduction zone offshore in the Ionian Sea (Fig. 6.7).

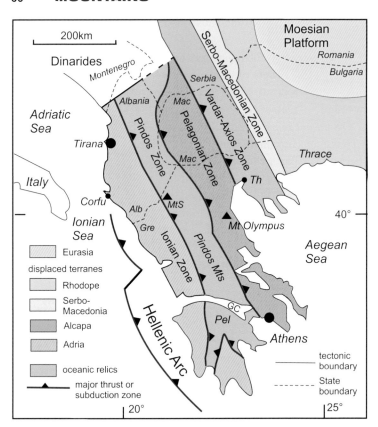

Figure 6.7 Tectonic Zones of the Hellenides. Alb, Albania; Bulg, Bulgaria; GC, Gulf of Corinth; Gre, Greece; Mac, Macedonia; MtS, Mt Smolikos; Pel, Peloponnese; S, Sophia; Ser, Serbia; Th, Thessaloniki. After Degnan & Robertson, 2006.

Figure 6.8 Sandstone cliffs in the Meteora Mountains: Pindos Range, Thessaly, Greece. © Shutterstock, by Anton Ivanov.

Eight tectonic zones are distinguished in this sector of the orogenic belt which, including the Rhodope Massif, has widened here to around 1200km. These are, from north to south: the Moesian Platform (part of the Eurasian foreland); the Rhodope Massif; the Serbo-Macedonian Zone; the Vardar–Axios Zone; the Pelagonian Zone; the Pindos (or Sub-Pelagonian) Zone; the Ionian Zone and the Hellenic Arc.

The Moesian Platform

This region, forming the southern foreland to the Carpathians, although belonging to the Eurasian continental plate, is now separated from it by a narrow Triassic rift. It consists of Palaeozoic basement with a Mesozoic shallow-marine sedimentary cover.

The Rhodope Massif

This broad zone consists of a high-grade metamorphic basement of Palaeozoic age together with a lower-grade metamorphosed Mesozoic cover, which includes a late Jurassic–early Cretaceous subduction–accretion complex. These metamorphic rocks are overlain unconformably by a late Cretaceous to early Neogene sedimentary cover. The Massif has experienced a late Cretaceous compressional event, which produced a stack of thrust nappes, and also an extensional event in the mid-Palaeogene that reversed the sense of movement on some of the earlier thrusts.

The Massif is considered to be part of the same terrane as Dacia, which forms part of the basement of the Pannonian Basin and was discussed earlier in the context of the Carpathians (see Fig. 6.2). This terrane was separated from the rest of the Eurasian continent by the early Mesozoic Vardar oceanic basin; this closed during the early Cretaceous Eo-Alpine collision that formed the Eastern Thrust Belt of the Dinarides.

The Serbo-Macedonian Zone

This zone is a Palaeozoic metamorphic massif, superficially similar to the basement of the Rhodope Massif, but exhibiting significant differences in its geological history. It is considered to be a separate terrane that joined the Rhodope Massif during the Hellenide orogeny.

The Vardar–Axios Zone

This zone consists of oceanic material of Mesozoic to Palaeogene age, including ophiolites, and is believed to represent oceanic crust of the Vardar Ocean that formerly separated the Gondwanan terranes from the Eurasian continental margin.

The Pelagonian Zone

This represents a terrane, originally part of the Alcapa Terrane, believed to have originated from Gondwana, and to have been separated from Eurasia by the Vardar Ocean (see Fig. 6.6). It appears to be the westwards continuation across the Aegean Sea of the Anatolian Terrane of Turkey (see Fig. 6.1).

The Pindos (or Sub-Pelagonian) Zone

This is similar to the Vardar–Axios Zone in containing oceanic material, but is somewhat older; it has been interpreted as an oceanic rift basin of Triassic age (the 'Pindos Ocean') that originally separated the Pelagonian Terrane from Gondwana. It now consists of a stack of folded thrust nappes containing a Triassic to Palaeogene sedimentary sequence, directed south-westwards over the underlying Ionian Zone.

The Ionian Zone

This zone, also known as the Gavrovo-Tripolitsa Zone, is the southern continuation of the Adrian platform in the Dinarides to the northwest. It forms the southwestern foreland onto which the Pindos nappes have been overthrust.

The Hellenic Arc

This currently active subduction zone begins just south of Corfu and curves around the southern side of Crete (see Fig. 6.1), forming the outer margin of the Hellenide sector of the Alpine Orogenic Belt. The Hellenic Arc is discussed in more detail below.

Tectonic overview

The Hellenide–Rhodope belt is a good example of a long-lived subduction–accretion complex, having formed as a result of repeated additions of both continental and oceanic terranes over a long period of time, beginning in the Jurassic. It includes the remnants of two major ocean basins – the Vardar Ocean in the northeast and the Pindos Ocean in the southwest (see Fig. 6.9).

The Rhodope Massif in the north became separated from the rest of the Eurasian continent by an early Mesozoic back-arc basin that formed on the upper plate of the subduction zone bounding the north side of the Tethys Ocean. This ocean basin was closed during the early Cretaceous (Eo-Alpine) contractional event that resulted in the Eastern Thrust Belt of the Dinarides.

The Vardar–Axios Zone represents another ocean basin, the Vardar Ocean, which was initiated in the

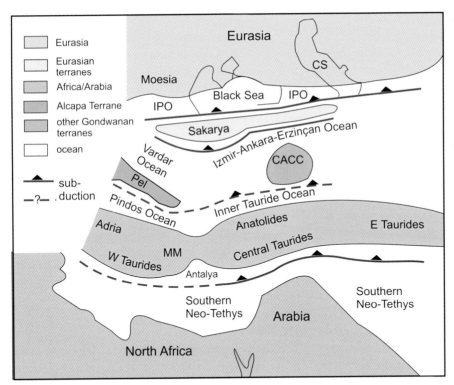

Figure 6.9 Arrangement of continent and ocean in the Eastern Mediterranean in the late Cretaceous: a speculative reconstruction. CS, Caspian Sea; IPO, Inner Pontide Ocean; MM, Menderes Massif; CACC, Central Anatolian Crystalline Complex. After Robertson, *et al.*, 2009

early Jurassic and existed through the Jurassic along the southern margin of the Eurasian continent. During this period, Vardar oceanic crust was being subducted north-eastwards beneath the Eurasian Plate. By the early Cretaceous, both the Rhodope Massif and Serbo-Macedonia had been added to Eurasia, and subduction continued beneath the Serbo-Macedonian Terrane. The Pelagonian Terrane, which had become detached from Alcapa, was situated across the Vardar Ocean to the southwest, and continued subduction through the Cretaceous led eventually to collision between Pelagonia and Serbo-Macedonia in latest Cretaceous to early Palaeogene times, producing the SW-directed Vardar–Axios fold-thrust belt.

The development of the Pindos Zone commenced in the Triassic as an oceanic basin offshore from the northeast margin of the Adrian Terrane. This ocean basin persisted throughout the Jurassic and early Cretaceous before being subducted beneath the Pelagonian Terrane in the mid-Cretaceous. The subduction complex is represented by an accretionary unit consisting of a stack of thrust sheets containing turbiditic flysch of early Palaeogene age. The main contractional event affecting the Pindos Zone was the development of a fold-thrust complex directed south-westwards towards the Ionian Platform during the mid-Palaeogene, which

spread throughout the whole Hellenide belt, welding together all the formerly separate terranes – Rhodope, Serbo-Macedonia, Pelagonia and Ionia-Adria.

The Hellenic Arc

The Hellenic Arc is the most tectonically active part of the Hellenides. It consists of an inner volcanically active island arc and an outer non-volcanic arc, which includes the large islands of Crete and Rhodes, together with several smaller islands (Fig. 6.10). The arc is seismically active and has been responsible for many large (magnitude 7) earthquakes over the last century. The volcanic arc includes five volcanoes that have been active historically: Methana, Milos, Nisyros and Santorini (two), together with a further three (Kos, Poros and Yali) active in the late Neogene and Quaternary Periods. Santorini is famous for the catastrophic explosion in 1470BCE, which has been held responsible for the destruction of the Minoan civilisation of Crete. The outer non-volcanic arc extends from the island of Lefkada, at the northern end of the Ionian Island chain, to the island of Rhodes, off the Turkish coast.

The outer non-volcanic arc is fringed on its outward side by an ocean trench, the Hellenic Trench, which is considered to be a fore-arc basin, developed between the

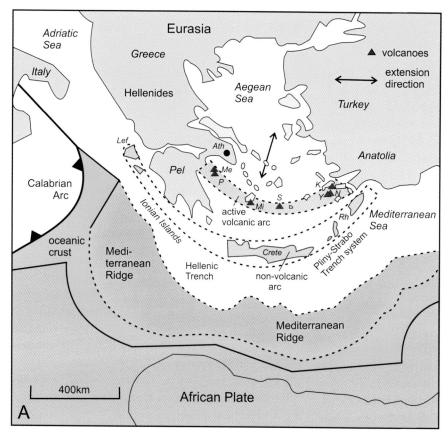

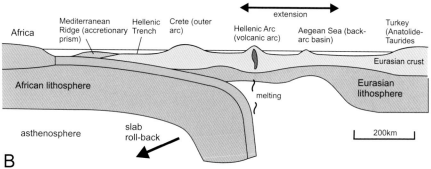

Figure 6.10 The Hellenic Arc System. **A**. Principal tectonic features: active volcanic arc (red); inactive outer arc (blue) and Mediterranean Ridge (accretionary prism) (orange). Colours of other tectonic units as for previous figures. The site of the subduction zone is concealed beneath the ridge. The southern part of the Aegean Sea is a back-arc basin. Active or recently active volcanoes: K, Kos; M, Methana; Mi, Milos; N, Nisyros; P, Poros; S, Santorini; Y, Yali. Other locations: Ath, Athens; Lef, Lefkada; Pel, Peloponnese; Rh, Rhodes. **B**. Schematic crustal profile across the central part of the Hellenic Arc showing the main tectonic elements. Vertical scale approx. x2. A, after Okay, 2000; B, after Dilek & Altunkaynak, 2009.

trench and the volcanic arc. The site of the subduction zone is concealed beneath the Mediterranean Ridge, which occupies a broad zone in the central part of the eastern Mediterranean Sea. This ridge is interpreted as an accretionary prism that has accumulated on the site of the subduction zone over a long period of time, filling in the original ocean trench. As can be seen in Figure 6.6, subduction of the oceanic part of the African plate has reached an advanced stage and only a small area remains south of the ridge.

The southern part of the Aegean Sea, inward of the volcanic arc, represents a back-arc basin, and the study of the active extensional faults, both in the Aegean islands and on the southern Greek mainland, indicates that the Hellenic Arc as a whole has experienced extension in an approximately NNE–SSW direction perpendicular to the arc, and also parallel to the arc as it has expanded outwards and stretched. The outward migration of the subduction zone, which has occurred over a period of at least 50Ma, from the Eocene Epoch to the present, has resulted in crustal thinning, leading to the depression of the whole Aegean Sea Basin. This process, which is common to many present-day island arc systems, is attributed to the gradual outwards retreat of the subduction zone due to the process of trench, or slab, roll-back, illustrated in Figure 6.10B.

Turkey

The Alpine mountain belt continues eastwards from Greece, crossing the Aegean Sea into Turkey (Figs. 6.1, 6.11). Almost the whole of Turkey consists of a series of mountain ranges surrounding the high plateau of Anatolia (Fig. 6.4). The northern branch of the mountain system, known as the Pontides, forms two arcs fringing the southern shore of the Black Sea. The shorter Western Pontide chain extends from east of Istanbul to the Sinop Peninsula on the Black Sea coast; the longer, more easterly, Central and Eastern Pontide chains form a great southward-convex arc, extending for about 650km from Sinop to the border of Georgia at Batum (Fig. 6.12). These mountains increase in height towards the east, the highest peak being at 3937m. The central part of western Turkey consists of the high Anatolian Plateau, over 1000m in height, on which the capital, Ankara, sits. This great plateau is about 700km in length and up to 300km across and separates the northern Pontide ranges from the Taurus ranges in the south.

The southern mountain system is known as the Taurides which, at its western end, form a north-pointing 'V' shape enclosing the Bay of Antalya. Further east, the range curves around the south of the Anatolian Plateau and extends in a northeasterly direction to meet the Pontides in the high mountainous region of eastern Turkey, around the city of Erzerum. Mt. Ararat, at 5165m the highest mountain in Turkey (Fig. 6.13) is located here, near the borders with Armenia and Iran.

The mountain belt then turns south-eastwards through Iran, to form the Zagros and Elburz ranges, discussed in the next chapter.

Tectonic setting

In geological terms, Turkey is a land of two halves: the western part, sandwiched between the Black Sea and the Eastern Mediterranean, is essentially a subduction–accretion complex, where subduction continues along the Hellenic and Cyprus trenches (see Fig. 6.1), whereas the eastern half is a collision belt between the Eurasian Plate in the north and the Arabian Plate in the south.

The main Alpine orogenic belt in Western Turkey (Fig. 6.12) is over 600km across at its widest point – over 800km if Cyprus is included – and is made up of two main tectonic zones, the Pontide Zone in the north and the Anatolian–Tauride Zone in the south, separated by the Izmir–Ankara Suture. The northern boundary of the Pontides is a major thrust that lies offshore along the coast of the extensional Black Sea Basin. The southern boundary of the Anatolian–Tauride Zone is effectively the Hellenic and Cyprus Trenches, which mark the currently active subduction zone defining the margin of the oceanic part of the African Plate.

In Eastern Turkey, the orogenic belt widens to include the Caucasus ranges in the north, and is bounded in the south by the Assyrian–Zagros suture marking the northern edge of the Arabian Plate. In this sector, the belt as a whole is just over 600km across. Here there are four

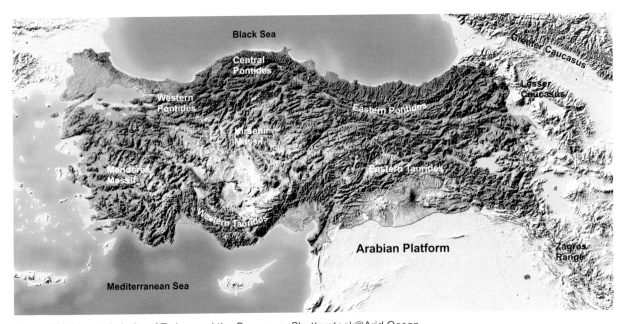

Figure 6.11 Mountain belts of Turkey and the Caucasus. Shutterstock©Arid Ocean.

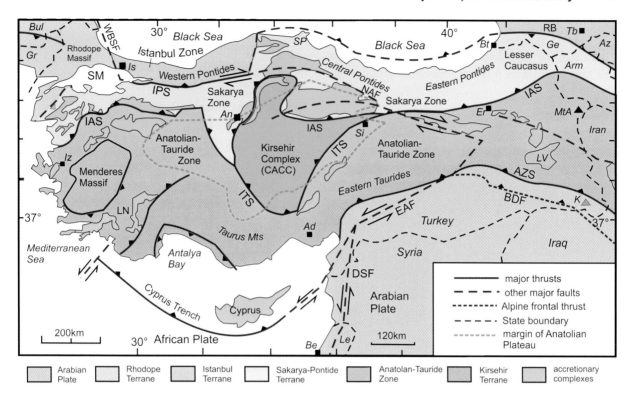

Figure 6.12 Main structural elements of the Turkish Alpine Belt. Note the distribution of accretionary complexes, including ophiolites, in the Sakarya and Anatolide–Tauride Zones. AZS, Assyrian–Zagros Suture; BDF, Bitlis deformation front; CACC, Central Anatolian crystalline complex; DSF, Dead Sea Fault; EAF, East Anatolian Fault; IAS, Izmir–Ankara Suture; IPS, Intra-Pontide Suture; ITS, Inner Tauride Suture; LN, Lycian Nappes; NAF, North Anatolian Fault; RB, Rioni Basin; WBSF, West Black Sea Fault. Countries: Arm, Armenia; Az, Azerbaijan; Bul, Bulgaria; Ge, Georgia; Gr, Greece; Le, Lebanon. Cities: Ad, Adana; An, Ankara; Er, Erzerum; Is, Istanbul; Iz, Izmir; Si, Sivas; Tb, Tbilisi. Geographic names: LV, Lake Van; SM, Sea of Marmara; SP, Sinop Peninsula. After Okay, 2000.

Figure 6.13 Mount Ararat, the highest mountain in Turkey. This is an extinct volcano situated on the high plateau near the Armenian border and the traditional resting place of Noah's Ark. © Shutterstock, by Alexander Ishchenko.

main tectonic zones: the Greater Caucasus and Lesser Caucasus in the north, separated by the Rioni–Kura Basin, and the Anatolian–Tauride Zone in the south. The Lesser Caucasus is the eastern continuation of the Pontide Zone.

The Pontide Zone

The Pontides are an amalgamation of three separate terranes: the Strandja, Istanbul and Sakarya Terranes, which came together during the mid-Cretaceous. The Strandja Terrane is the eastern equivalent of the Rhodope Massif already referred to when discussing the Hellenides (see Fig. 6.7), and forms the basement of that part of Turkey northwest of the Sea of Marmara. It is cut off on its eastern side by a NNW–SSE transform fault, the West Black Sea Fault, which terminates the extensional West Black Sea Basin. East of this fault is the Istanbul Terrane, considered to have been part of the Moesian Platform, now on the northwest side of the Black Sea, until the early Neogene, when it was displaced southwards by the opening of the Black Sea Basin.

Like the Rhodope Massif, the Strandja and Istanbul terranes have a Palaeozoic basement of Laurasian type. These terranes are separated from the Sakarya Terrane by the NW-dipping Intra-Pontide Suture. This suture zone consists of imbricated thrust units of Mesozoic oceanic material considered to represent the remnants of the Neo-Tethyan Vardar Ocean, which was subducted beneath the Istanbul Terrane until the two terranes collided in the Palaeogene.

The Sakarya Terrane occupies the remainder of the Pontide Zone from the Mediterranean coast north of Izmir, eastwards along the southern side of the Black Sea towards the border with Georgia. It has a Palaeozoic basement of Eurasian parentage, with a Jurassic to mid-Cretaceous platform cover. The Sakarya Terrane is a composite unit containing numerous thrust slices, some of which contain Cretaceous flysch deposits and oceanic material, including ophiolites. The northern boundary of the Sakarya Terrane in the Central and Eastern Pontides is concealed beneath the Black Sea Basin but appears in the Caucasus as a south-dipping thrust defining the northern margin of the Lesser Caucasus. The Caucasus sector is discussed separately below. The southern boundary of the terrane is the Izmir–Ankara suture zone, which is another complex thrust belt like the Intra-Pontide suture, containing oceanic material attributed to the Vardar Ocean. The western boundary of the terrane is obscured beneath the Aegean Sea.

All three terranes in the Pontide Zone have experienced significant Cretaceous deformation as well as the main late Palaeogene to early Neogene Alpine tectonic event.

The Anatolian–Tauride Zone

This zone occupies the southern part of Turkey, bounded in the north by the Izmir–Ankara Suture, in the southwest by the Mediterranean Sea and in the southeast by the Assyrian-Zagros Suture. It differs from the northern terranes in possessing a basement of African type, similar to that of the Arabian Plate, affected by early Cambrian (Pan-African) and late Palaeozoic orogenic events. There are three main components in this zone: the Taurides, the Kirsehir Massif (also known as the Central Anatolian Crystalline Complex or CACC) and the Menderes Massif, the latter two being known collectively as the Anatolides. The Anatolides represent the metamorphosed, structurally lower, northern part of the zone and would have been underthrust beneath the Sakarya Terrane during the collision with it.

The Taurides consist of a series of thrust sheets containing mostly sedimentary rocks of Palaeozoic to early Neogene age. The top thrust slice in each unit is composed of ophiolitic material which forms large outcrops throughout the Taurides, some of the more prominent of which are shown in Figure 6.12. The thrust sheets were mostly assembled during the late Cretaceous and culminated in the early Palaeogene with a major contractional event caused by the collision between the Anatolian–Tauride block and the Pontides. The thrust sheets were emplaced southwards with the exception of the southernmost, around the Bay of Antalya, which appear to have been emplaced towards the north. The more northerly thrust slices, which belong to the lower plate in the collision with the Pontides, and were overthrust by them, have been subjected to high-pressure–low temperature metamorphism. The Lycian nappes in the south are overlain by ophiolites considered to represent the southern branch of the Neo-Tethys Ocean.

The Central Anatolian Crystalline Complex (CACC), otherwise known as the Kirsehir Massif, occupies the central part of the Anatolian Plateau. It consists of metamorphic and granitic rocks of late Cretaceous age overlain by Neogene sediments and volcanics. The complex is bounded in the north by the Izmir–Ankara suture separating it from the Pontides, and in the south by the Inner Tauride Suture, separating it from the other units of the Anatolian–Tauride Zone. Both thrust boundaries are inclined northwards, and, like the Tau-rides, the complex lay on the lower plate of the collision with the Pontides. The complex may represent an individual terrane originally isolated from the Pontide and Tauride blocks by branches of the Neo-Tethys Ocean, as indicated in Figure 6.9.

The Menderes Massif is located in the westernmost part of the Anatolide–Pontide Zone, southeast of the city of Izmir. It has experienced a similar tectonic history to that of the Kirsehir Massif and has been described as a metamorphic 'core complex', similar to other Alpine core complexes, such as the Alboran and Pennine zones discussed in previous chapters, in possibly having been uplifted during an extensional process. However, unlike the Kirsehir Massif, it may not represent a separate terrane. The Menderes Massif is bounded on its southern side by a south-dipping thrust. The Tauride units southeast of this boundary, known as the Lycian Nappes, consist of unmetamorphosed units overthrust westwards onto the Massif.

The southern margin of the Anatolian–Tauride Zone is the plate boundary marking the northern edge of the African Plate. It consists of two arcuate trenches, the Hellenic Trench in the west, shown in Figure 6.10, and the Cyprus trench in the centre, connected by a transform fault. The eastern end of the Cyprus trench is linked via another transform fault to the Dead Sea Transform Fault and the Assyrian–Zagros Suture, which together form the boundary of the Arabian Plate. Cyprus itself contains the famous Troodos Massif, one of the first ophiolite complexes to be recognised.

Tectonic history

At the beginning of the Mesozoic, much of present-day Turkey lay on the northern margin of Gondwana, separated from Eurasia by the wide Tethys Ocean (see Fig. 4.4A). Oceanic lithosphere of the Palaeo-Tethys Ocean was being subducted southwards beneath the northern margin of Gondwana, and northwards beneath Laurasia, resulting in the opening of a series of back-arc basins in the Triassic. These new oceanic basins, collectively known as Neo-Tethys, separated the new continental terranes from both Laurasia and Gondwana. The northern Neo-Tethys branches are now represented by the Intra-Pontide, Izmir–Ankara and Inner Tauride sutures. The southern branch of Neo-Tethys is represented by the remnant of the oceanic part of the African Plate, which is being subducted beneath the Cyprus Trench. During the Jurassic, the original Tethys Ocean (Palaeo-Tethys)

was consumed, leaving only the various branches of Neo-Tethys.

Figure 6.9 is a reconstruction of the possible arrangement of continental and oceanic areas in the late Cretaceous. The Sakarya continent lay within the northern Neo-Tethys, separated from Eurasia by the Intra-Pontide Ocean. South of this were two other small terranes, the Pelagonian and Kirsehir Terranes, also within the northern Neo-Tethys Ocean. Across the Aegean Sea, in the Hellenides, the Pelagonian Terrane divided this large oceanic area into two branches: the Vardar Ocean in the north and the Pindos Ocean to the south; however, this terrane does not appear to extend eastwards into Turkey. Here, the northern Neo-Tethys is known as the Izmir–Ankara Ocean in the north and the Inner Tauride Ocean in the south. All these oceanic areas were probably connected. The southern margin of this Neo-Tethyan Ocean lay along the northern edge of the large Anatolian–Tauride Platform, which was probably continuous with the Adria–Alcapa platform further west in the Balkans, and represented by the Ionian Zone in the Hellenides (see Figs 6.5 and 6.7). All these terranes were covered by shallow-marine sediments.

The Anatolian–Tauride Massif is part of a large super-terrane named 'Cimmeria' which was attached to Gondwana until the beginning of the Triassic Period, when it broke away, opening up the southern branch of the Neo-Tethys Ocean. Cimmeria was a long, thin micro-plate, parts of which now form the Iranian, Afghanistan and Tibetan Plateaux in addition to the Anatolian Plateau, and which also extends eastwards into Indochina and Malaysia.

In the south, where the southern Neo-Tethys Ocean lay between the Anatolian-Tauride platform and Africa, the small Cyprus terrane became separated from the southern margin of Anatolian-Taurides by the Antalya back-arc basin.

From mid-Cretaceous times, consumption of the Neo-Tethyan oceanic lithosphere occurred by northwards subduction beneath the Eurasian, Sakaryan and Anatolian-Tauride margins. The Black Sea basin developed as a back-arc basin above the southern Eurasian margin, splitting off the Istanbul terrane from the rest of Eurasia. In the late Cretaceous, ophiolite nappes were being obducted onto the northern margins of the Anatolian-Tauride and Arabian Platforms. During the Eocene, the Anatolian-Tauride Platform collided with Sakarya, Strandja–Istanbul and southern Eurasia to create the Pontides; and by the Miocene, Arabia had collided with the enlarged Eurasian Plate, producing strong contractional deformation in the Central and Eastern Taurides and also throughout the whole system. This Miocene event produced southward-directed thrust slices, causing the imbrication of the Anatolian-Tauride platform. Since then, the main tectonic movements have been concentrated along a major strike-slip fault system based on the North and East Anatolian Faults, which have accommodated the further convergence between the Arabian and Eurasian Plates by westwards movement of the Anatolian Plateau.

Figure 6.14 is a reconstruction of the possible sequence of tectonic events leading up to the present disposition of terranes. There are other possible ways of explaining the present geology, and the actual history is very much more complex, but the sequence shown here is an illustration of how a subduction–accretion complex can be built up from successive events of extensional rifting, subduction, and collision.

The Caucasus

The Caucasus sector of the Alpine Orogenic Belt consists of two distinct mountain belts: the Greater and Lesser Caucasus, separated by the Trans-Caucasian Massif. As described earlier, the Lesser Caucasus is the lateral extension of the Pontide Zone of northern Turkey (see Fig. 6.12). The western end of the Greater Caucasus is cut off by the Black Sea, but reappears in the Crimean Peninsula as the Yayla Range, itself terminated by the West Black Sea Basin at its western end (see Fig. 6.1). However, the southern Eurasian margin continues westwards along the southern margin of the Black Sea to link up with the Balkanide range, which defines the

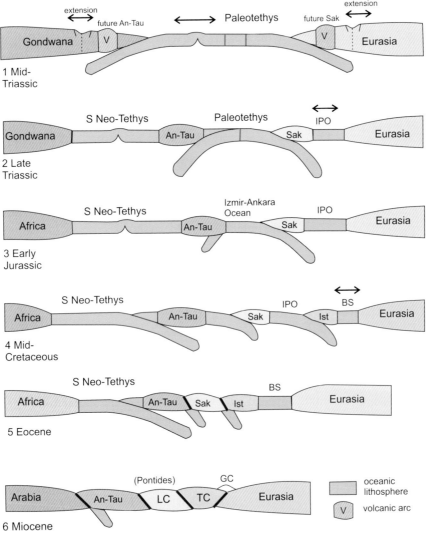

Figure 6.14 Possible sequence of tectonic events affecting the Turkish sector of the Alpine belt. Cartoon sections, greatly simplified: 1–5 Western Turkey; 6, Eastern Turkey and the Caucasus. An-Tau, Anatolian–Tauride Terrane; BS, Black Sea Basin; GC, Greater Caucasus; IPO, Intra-Pontide Ocean; Ist, Istanbul Terrane; LC, Lesser Caucasus (= Sakarya Terrane); TC, Trans-Caucasian Terrane.

southern limit of the Moesian Platform. The eastern end of the Greater Caucasus range is cut off at the Caspian Sea coast, but the Lesser Caucasus curves around the southern end of the Caspian Sea to join the Elburz Range in Iran, discussed in the following chapter.

The Greater Caucasus mountain range extends for *c*.1300km from the northeastern shore of the Black Sea to the western shore of the Caspian Sea (Figs 6.15, 6.16). It includes many mountains over 5000m high, including Mount Elbrus which, at 5641m, is Europe's highest mountain. Many of these peaks are Neogene volcanoes – some, including Mount Elbrus itself, recently active, and the region is strongly seismic, having experienced many severe earthquakes in recent times. The Yayla sector is narrow and relatively insignificant in comparison with the Caucasus ranges; the Balkanides, however, although also narrow, contain several prominent mountains, including Yamruchka, the highest peak in the range, at 2376m.

The Greater Caucasus Zone

This zone is divided into four tectonic units: the Scythian Platform, or northern foreland, the Front Range, Main Range, and Southern Slope (Fig. 6.15).

The Scythian Platform has a Palaeozoic basement of Eurasian origin, mostly concealed under a thick cover of Cenozoic molasse. The molasse deposits are concentrated in separate foredeep basins immediately north of the Caucasian Front, including the Azov–Kuban, and Terek–Caspian Basins, which are separated by basement highs.

The Greater Caucasus Front Range consists of a stack of north-directed thrust sheets containing Palaeozoic basement overlain by an unconformable cover of Jurassic sediments and volcanics. The Greater Caucasus Main Range also consists of a Palaeozoic basement complex folded into a series of antiforms separated by synforms containing Jurassic black slates. This zone is thrust southwards over the Southern Slope Zone.

The Southern Slope unit consists of a similar sequence to the Front and Main Ranges overlain unconformably by late Jurassic to early Palaeogene flysch deposits. This complex was intensely folded in the early Neogene and affected by south-directed overthrusts, the most southerly of which separates it from the Trans-Caucasian Massif.

The Trans-Caucasian Massif

This zone forms a topographic depression between the Greater and Lesser Caucasus mountain ranges. It consists of a late Palaeozoic (Variscan) basement overlain by a Mesozoic to Cenozoic cover of sediments and volcanics.

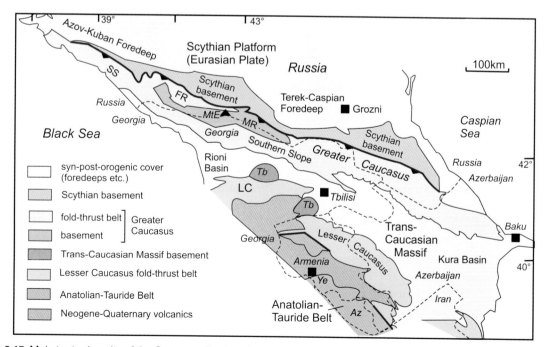

Figure 6.15 Main tectonic units of the Caucasus Region. Greater Caucasus: FR, Front Range; MR, Main Range; SS, Southern Slope. LC, Lesser Caucasus; Tb, Trans-Caucasian basement. Az, Azerbaijan; MtE, Mt Elbruz; Ye, Yerevan. After Adamia *et al.*, 2011.

Figure 6.16 The Caucasus Mountains. The Greater Caucasus Range on the border between Russia and Georgia is the result of the final collision between the Arabian and Eurasian Plates. © Shutterstock, by Lizard.

The pre-Mesozoic basement is exposed in the central part of the zone but is overlain by sedimentary basins, the Rioni Basin in the west, and the Kura Basin in the east. The Trans-Caucasian Terrane is considered to have separated from Eurasia during the early Mesozoic by the formation of a back-arc basin, then collided with Eurasia again in the early Neogene. In this respect it is similar to the Istanbul Terrane in Western Turkey.

The Lesser Caucasus

This zone is the eastwards extension of the Pontides of Turkey, discussed above.

Tectonic history

The Caucasus region has experienced repeated periods of orogenic activity (Fig. 6.14). The Trans-Caucasian Terrane appears to have separated from the Eurasian margin during the Triassic. The Lesser Caucasus Terrane, equivalent to the Pontides Zone further west, was added to the Eurasian margin in the early Jurassic and the Anatolian–Tauride Terrane in the Cretaceous, as well as pieces of oceanic and island-arc crust, as Neo-Tethyan oceanic crust was subducted northwards beneath Eurasia. The latest, and most severe, orogenic episode was caused by collision between the previously assembled terranes and the Arabian Plate during the late Miocene Epoch. Tectonic activity has continued to the present day: large earthquakes (magnitude 6–7) have recently affected the southern foothills of the Greater Caucasus, associated with the suture between it and the Trans-Caucasian Massif, and Neogene to Quaternary volcanic activity has produced the extensive lavas of the Armenian Plateau together with extinct volcanoes, such as Elbrus, in the Greater Caucasus.

7

Iran to Pakistan

Iran

The main Alpine mountain belt turns sharply south-eastwards from Eastern Turkey into Iran, where it divides into two branches. The more southerly becomes the Zagros Range, which continues along the Iraq–Iran border and the northeast side of the Persian Gulf to the Straits of Hormuz – a distance of around 1300km (Figs 7.1, 7.2). The Zagros Range is an impressive mountain belt, rivalling the Western Alps in scale, being 400km wide in the northwest to over 600km in the southeast. Much of it is over 3000m in height, with several peaks over 4000m, the highest being Zard Kuh (4548m) in the central part of the Zagros. The southeastern end of the Zagros Range ends abruptly at the Straits of Hormuz, and east of this is a narrow arcuate mountain range facing out towards the Gulf of Oman. This region is known as the Makran and extends eastwards into southwest Pakistan.

The northern branch of the Alpine belt in Iran is the Alborz Range, which curves around the southern side of the Caspian Sea, and is much narrower, only 60–130km wide, but is even higher and contains the highest mountain in Iran, Damavand, at 5604m. The eastern end of the Alborz Range merges into a broader belt of mountains, the Kopeh Dagh, that follow the northeastern border of Iran and extend into northern Afghanistan, where they eventually join the Hindu Kush.

The Zagros and Alborz mountain ranges are separated by the vast Central Iranian Plateau, which is roughly triangular in shape, broadening to nearly 400km width in the centre. The northwestern part of the plateau is known as the Dasht-e-Kavir, or 'Great Salt Desert', and the southern as the Dasht-e-Lut. The whole plateau is over 400m in height, is completely enclosed by mountains, and contains many seasonal salt lakes and marshlands. A third belt of mountains runs along the Iran–Afghanistan border, forming the eastern side of the Central Iranian Plateau.

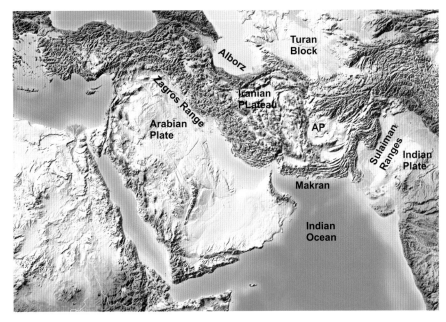

Figure 7.1 The Zagros and the Makran. Note that the Zagros Range bordering the Arabian Plate is separated from the Alborz Range in the north by the Central Iranian Plateau. The Turan Block north of the Alborz is part of the Eurasian Plate. AP, Afghan Plateau. © Shutterstock, by Arid Ocean.

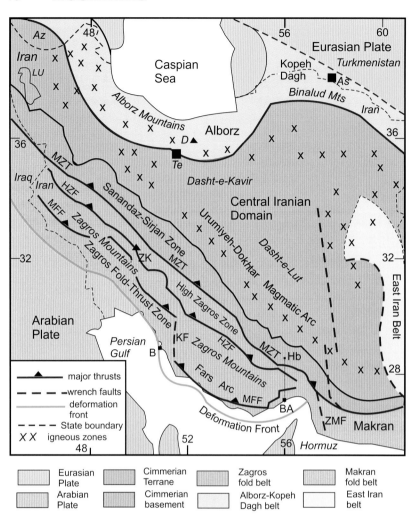

Figure 7.2 Major tectonic features of Iran. HZF, High Zagros Fault; KF, Kazerun Fault; MFF, Main Frontal Flexure; MZT, Main Zagros Thrust; ZMF, Zendan-Minab Fault. As, Ashkhabad; Az, Azerbaijan; B, Bushehr; BA, Bandar Abbas; Hb, Hajiabad; Te, Tehran. D, Damavand; ZKV, Zard Kuh. After Paul *et al.*, 2010.

The Zagros Belt

Compared to most of the Alpine belts discussed so far, the Zagros belt is comparatively simple in its basic structure. It is the product of the collision between the Arabian Plate in the southwest and the Iranian Terrane, or microplate, in the northeast. There are three tectonic zones: the Zagros Fold-Thrust Zone on the southwest side of the belt, the High Zagros Zone in the centre, and the Sanandaz–Sirjan, or Internal, Zone to the northeast (Figs 7.2, 7.3).

The stratigraphic sequence in the Zagros belt consists of between 6 and 15km of sedimentary strata overlying a Precambrian metamorphic basement. The sedimentary sequence consists of Palaeozoic, mainly clastic, sediments overlain by Mesozoic to Palaeogene massive carbonates, then finally Miocene to Recent molasse. At the base of the sequence in the southeast is a thick layer of salt, the Hormuz Salt, which provides an important décollement surface for the larger folds and is the dominant factor controlling the fold style in the southern part of the Fold-Thrust Zone (Fig. 7.3).

The Zagros Fold-Thrust Zone

This zone is a typical thin-skinned fold-thrust belt, affecting mainly the sedimentary platform cover of the passive margin of the Arabian Plate. The zone is bounded to the northeast by the High Zagros Fault, which separates it from the High Zagros Zone. The outer margin of the Zagros Fold-Thrust Zone is defined by the Zagros Deformation Front, which lies some distance within the Arabian Platform in the northern sector and continues offshore within the Persian Gulf in the south. Between these two lineaments is the Mountain Front Flexure, which defines the outer margin of the more strongly folded part of the zone and marks the edge of the coastal plain in the north. The Fold-Thrust Zone is divided into two

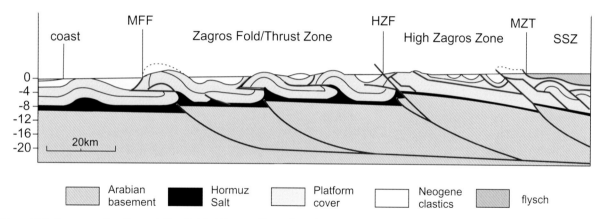

Figure 7.3 Cross-section through the Folded Zagros Zone, from Bandar Abbas to Hajiabad. Note the role of the Hormuz salt as a décollement layer. The size of the folds is dependent on the thickness of the platform cover, which consists of c.4km of Cambrian to Eocene sediments dominated by carbonate. The section is 160km long. HZF, High Zagros Fault; MFF, Mountain Front Flexure; MZT, Main Zagros Thrust; SSZ, Sanandaj-Sirjan Zone. After Regard *et al.*, 2010.

sectors by the N–S Kazerun Fault, a dextral wrench fault, which displaces the Mountain Front Flexure by about 150km, from the northeastern margin of the coastal plain to the coast at Bushehr.

The southern sector of the Fold-Thrust Zone, known as the Fars Arc, faces south towards the Persian Gulf and contains some of the more spectacular fold structures of the Zagros Belt. When flying over this section of the Zagros by commercial airline, it is possible to see the individual anticlinal structures clearly, as they form prominent mountain ridges. As illustrated in Figure 7.4, these anticlines are actually anticlinal periclines, shaped like upturned canoes, and only persist for relatively short distances (e.g. 20–30km) along strike. Figure 7.3 is a cross-section through part of the Fold-Thrust Zone, showing how the shape and size of the folds are controlled by the thickness of the carbonate sequence above the basal salt layer, which acts as a décollement surface.

The High Zagros Zone

This consists of highly deformed Mesozoic metamorphic rocks and is bounded to the northeast by the Main Zagros Thrust, more precisely a reverse fault, which has been shown from seismic evidence to dip northeastwards beneath the Central Iranian Massif, and represents the plate boundary between the Arabian and Iranian plates. The zone is bounded in the southwest by the High Zagros Fault, which separates it from the Fold-Thrust Zone. The highest part of the mountain belt, which includes Zard Kuh, lies immediately northeast of this fault.

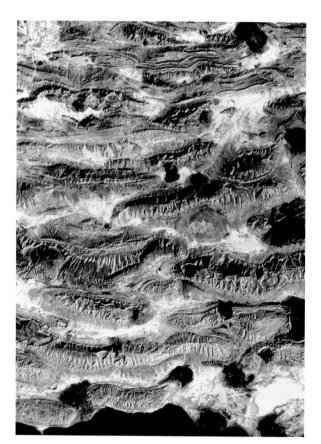

Figure 7.4 Periclinal anticlines in the Fars Arc, southern Zagros Mountains. This is a thermal image. The mountain ridges (in red) are formed along the crests of anticlines, exposing carbonate strata, the red colour indicating the presence of vegetation; the valleys (paler colour) are filled with younger sediments and are more arid. The coast of the Persian Gulf is visible at the bottom of the image. The folds are several tens of kilometres long and around 10km across.
© Science Photo Library

The Sanandaz–Sirjan Zone

This is the internal metamorphic complex of the orogenic belt and consists of an outer series of thrust slices containing deep-marine sediments, ophiolites and volcanics, and an inner belt of highly deformed late Palaeozoic to Mesozoic platform sediments originating on the passive margin of the Iranian Terrane. The zone is bounded on its northeast side by the Central Iranian Massif and represents that part of the Massif that lies within the Zagros tectonic belt and has been affected by the Alpine deformation.

Tectonic history

Rifting beginning in the late Permian to early Triassic resulted in the separation of the Iranian Terrane, part of the large composite Cimmerian Terrane, from the Arabian margin of Gondwana. Cimmeria moved northwards towards Eurasia during the Jurassic, driven by subduction of the Palaeo-Tethyan Ocean crust, and finally collided with the southern boundary of Eurasia during the late Triassic to early Jurassic.

Subduction of the Neo-Tethyan Ocean crust beneath the Iranian Terrane took place during the late Jurassic and Cretaceous Periods and reached a maximum in the late Cretaceous, resulting in the formation of a magmatic arc along the southwest edge of the Central Iranian Massif. Final collision between Arabia and the Iranian Terrane occurred in the Miocene Epoch, and convergence is still taking place at a rate of 5–10mm/a, accompanied by many major earthquakes. It has been estimated that about 59km of shortening (23%) has taken place to date across the southern part of the Zagros.

The Central Iranian Plateau

The vast region of the Central Iranian Plateau, measuring about 1100km in length and 400km across at its widest part, is completely surrounded by mountains: the Zagros in the southwest, the Alborz and Kopeh Dagh in the north, and the East Iran Belt in the east. Although much of the Plateau consists of flat arid plains with many seasonal salt lakes, there are considerable mountainous areas as well, including the peak of Kuh-e Darband, at 2499m.

In geological terms, the plateau, or Central Iranian Massif, is part of the large composite Cimmerian Terrane (Cimmeria), which is the south-eastwards continuation of the Anatolian–Tauride Zone discussed in the previous chapter, and is thought to have separated from Gondwana in the late Permian to early Triassic. It is composed of large Cenozoic sedimentary basins separated by uplifted massifs of metamorphosed basement rocks consisting of gneisses overlain by Mesozoic metasediments. The basins contain up to 4km of post-Eocene evaporites, carbonates and shales.

The Massif hosts large quantities of igneous material in two distinct categories: the earlier suite is mainly of Eocene age and is related to the north-eastwards subduction of the Neo-Tethyan ocean lithosphere; it is concentrated in the Urumiyeh–Dokhtar belt along the southwest margin of the plateau, bordering the Zagros Belt. The younger suite consists of post-collisional volcanics of Oligocene to Quaternary age scattered across the plateau.

Both the crust and mantle lithosphere of the plateau are much thinner than the adjoining mountain belts and exhibit a higher heat flow. These characteristics are attributed to a phase of mid-Miocene post-orogenic thinning and uplift, similar to that of other Alpine metamorphic core complexes, caused by the processes of trench roll-back and slab break-off described earlier.

The Northern Ranges
The Alborz

The Alborz Mountains form a well-defined, narrow, south-facing arc around the southern end of the Caspian Sea, and define the northeastern boundary of the Central Iranian Plateau. They are the Iranian continuation of the Lesser Caucasus Range of Armenia and Azerbaijan (see Fig. 6.15). The Alborz belt is mainly composed of a sedimentary sequence of Devonian to Oligocene age, dominated by Jurassic carbonates, and by late Cretaceous to Eocene volcanics resulting from the subduction of Tethyan oceanic lithosphere beneath the Iranian Terrane. This stratigraphic sequence was deformed during the Miocene collision between Iranian Terrane and the Arabian Plate, which continued into the Pliocene.

The Kopeh Dagh

The southern part of the Kopeh Dagh Mountains, known as the Binalud Range, is the eastern extension of the Alborz Range and forms the northeastern boundary of the Iranian Plateau, separating Iran from Turkmenistan to the north. The range extends in a southeasterly direction from the end of the Alborz to the border with Afghanistan. The northern section of the Kopeh Dagh strikes north-westwards towards the Caspian Sea coast at Krasnowordsk, and if this line is continued across the Caspian Sea to Baku, it reaches the southeast end of the Greater Caucasus. This NW–SE line thus marks the geological margin of the pre-Mesozoic Eurasian continent. There are three parallel ranges within the Kopeh Dagh, each containing peaks over 3000m in height, the highest, at 3314m, in the Binalud Mountains.

The Alpine tectonic history of the Alborz–Kopeh Dagh belt commenced with the Cimmerian Orogeny in the late Triassic to early Jurassic, which resulted from the closure of the Palaeo-Tethys Ocean. This was followed by a major extensional event in the mid-Jurassic producing a rift basin in which over 7km of Mesozoic sediments were deposited. The main orogenic phase, which caused the folding, thrusting and uplift of the fold belt, occurred in the late Eocene Epoch as a result of the Arabian–Iranian plate convergence further south in the Zagros. The major fold and thrust structures of the Kopeh Dagh are parallel to the suture defining the margin of Eurasia and are probably, in part, inherited from earlier Mesozoic structures. The last significant tectonic event occurred during the Pliocene, as the region was subjected to dextral wrench faulting due to the now oblique nature of the Arabian–Iranian convergence.

Tectonic summary

The Alpine orogenic belt of Iran differs in scale from the Mediterranean belts discussed in previous chapters, mainly because it resulted ultimately from the collision of two large continental plates. Consequently, whereas the Mediterranean belts have experienced histories that involved the subduction of Tethyan and Neo-Tethyan oceanic lithosphere, and the accretion of small terranes, the final collision of Africa and Eurasia has not yet occurred and there are still large marine areas within the overall belt underlain by oceanic or thinned continental crust. In contrast, the whole of the Iranian belt has experienced the effects of the final Arabian–Iranian collision, which has transmitted the contractional stress resulting from the Zagros collision through the Iranian terrane to the old Mesozoic suture zone at the Eurasian margin to form the northern fold belts. Slab roll-back at the Arabian margin has ceased, and the extensional region originally formed above the subduction zone is now an elevated plateau.

The Makran

The southern branch of the Alpine–Himalayan mountain belt continues eastwards from the Strait of Hormuz at Bandar Abbas along the northern side of the Gulf of Oman (Fig. 7.1). Here the mountain belt divides into two: a narrow northern arc, including two peaks over 3000m in height, forms the northern border of a central plateau, and a broader southern belt, known as the Makran, lies on the southern side. Topographically, the Makran belt in Iran consists of a high mountain ridge in the north, fringed by lesser hills to the south, sloping down towards the coast.

The Iranian sector of the Makran is about 550km long and extends eastwards along the north side of the Arabian Sea into the Baluchistan Province of Pakistan for a further 450km (Fig. 7.5). There are three arcuate mountain ranges within the Pakistani sector of the Makran Belt: the Siahan Range in the north, and the Makran Coast Range in the south, with the Central Makran Range between them. Only the latter two ranges include mountains over 2000m, and the belt as a whole is much less impressive than the Zagros.

Another mountain range strikes N–S along the border between Iran and Afghanistan, and sweeps round to join the Baluchistan ranges. This range, named the East Iran Belt in Figures 7.2 and 7.5, marks the boundary between the stable Iranian Massif in the west and the similar Afghan Terrane to the east in Afghanistan. A chain of late Cretaceous ophiolite outcrops and several major dextral wrench faults within this belt indicate there has been significant relative movement between the two stable blocks.

Plate-tectonic context

The western end of the arcuate Makran belt joins the eastern end of the Fars Arc, and the cusp where the two arcs meet, at the Strait of Hormuz, marks the point where the Arabian–Iranian plate boundary changes from a collisional suture to a subduction zone (Figs 7.1, 7.5). The Makran thus represents an accretionary complex resulting from the subduction of the oceanic part of the Arabian Plate beneath the Iranian and Afghan continental terranes to the north. A broad zone of dextral wrench faults, the Zendan–Minab fault system, links the end of the Main Zagros Thrust with the Makran Trench, which lies offshore in the Gulf of Oman. The individual structures of the Zagros Belt end at this lineament and are replaced by the separate fold-thrust structures of the Makran belt.

The Iranian Makran

The on-shore sector of the Iranian Makran occupies a 150km-wide belt between the Jaz Murian Depression

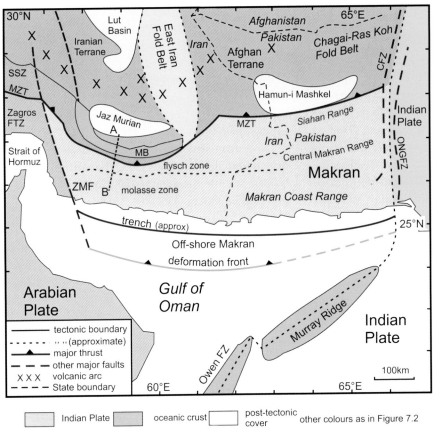

Figure 7.5 Main tectonic features of the Makran. CFZ, Chaman Fault zone; MB, marginal basin; MZT, Main Zagros Thrust; ONGFZ, Ornach-Nal Ghazaband Fault zone; Owen FZ, Owen Fracture Zone; ZMF, Zendan-Minab Fault zone; Zagros FTZ, Zagros Fold-thrust Zone. After McCall *et al.*, 1982.

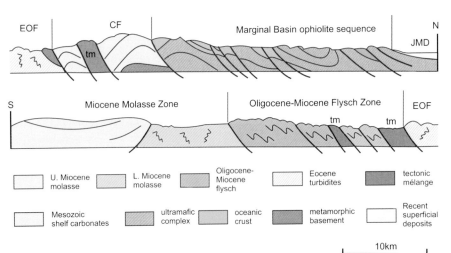

Figure 7.6 Structure of the onshore Makran. North–south section across the Iranian Makran along line A–B of Fig. 7.5 showing the main tectonic units. JMD, Jaz-Murian Depression; MB, Marginal Basin sequence; CF, Carbonate Fore-arc Zone; EOF, Eocene–Oligocene Flysch Zone; OMF, Oligocene–Miocene Flysch Zone; TM, tectonic mélange. After McCall *et al.*, 1982.

at the southern end of the Central Iranian Plateau and the coast, and can be divided into five tectonic zones (Figs 7.5, 7.6). From north to south, these are as follows.

1. The marginal basin, containing Jurassic to Palaeocene ophiolites and deep-marine sediments.
2. The carbonate fore-arc: a narrow zone of Palaeozoic metamorphic basement, known as the Bajgan-Durkan, the eastward continuation of the Sanandaz–Sirjan zone of the Zagros, with a Mesozoic platform cover.
3. The ophiolitic tectonic mélange zone, marking the continuation of the suture between the Arabian and Iranian plates (the Main Zagros Thrust).
4. The inner flysch zone, containing thick Eocene to Oligocene flysch.
5. The outer flysch zone, containing Oligocene to Miocene flysch.

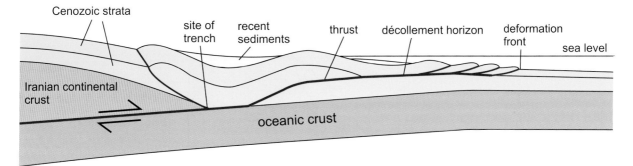

Figure 7.7 The Makran accretionary prism: a much simplified illustration of the structure. Note that the thrust-generated folding has progressed seawards from the original site of the trench and has also elevated the on-shore Makran. The thrusts have utilised a décollement horizon within the sedimentary sequence. After Platt *et al.*, 1985.

6 The molasse zone, containing Miocene to early Pliocene shallow-marine molasse.

The structure of zones 4–6 of the on-shore Makran, south of the suture zone, is dominated by steep north-dipping reverse faults separating south-facing overfolds, which is typical of a subduction–accretion complex.

The offshore sector of the Makran consists of a series of anticlinal ridges separated by narrow elongate basins (Fig. 7.7). The original topographic trench has been completely obscured by a 6–7km-thick pile of sediments that cover the oceanic crust. However, only the uppermost 2½km of sediment has apparently been involved in the folding that must overlie a décollement layer.

The subduction zone has been shown to dip north at a very shallow angle of *c.*10° and the subduction-related volcanic arc, consisting of a chain of Cenozoic volcanoes, lies 400–500km north of the coast. The estimated convergence rate is about 50mm/a, and the zone is still seismically active.

Tectonic History

The marginal (back-arc) basin now represented by the ophiolite complex developed on the southern side of the Central Iranian Plateau during the Jurassic, and remained active until the Palaeocene in response to the subduction of Arabian oceanic crust beneath the Iranian micro-plate. The development of this basin resulted in the isolation of the narrow Bajgan-Dur-kan basement terrane on its southern side. A subduction-related volcanic arc was created along the southern part of the Central Iranian Plateau.

The accretionary complex developed throughout the Cenozoic, forming two thick flysch sequences: Eocene to Oligocene, and Oligocene to Miocene, with an estimated total thickness of over 10,000m. A major compressional event in the late Miocene caused folding and thrusting throughout the on-shore Makran, including the closure and disruption of the back-arc basin, and the uplift of the Palaeogene flysch sequences. This event was probably related to the collision between the Arabian Plate and Iran in the Zagros Belt to the west, which also resulted in a change in convergence direction across the plate boundary.

Convergence continued in the Makran through the Pliocene and into the Pleistocene. The deformed Palaeogene sequences in the southern part of the zone were overlain by shallow-marine clastic sediments of molasse type. Sedimentation continued off-shore into the Gulf of Oman and was gradually added to the increasing accretionary prism. Deformation has continued to the present day in the offshore Makran as the deformation front, which is still active, migrated southwards to its present position about 150km from the coast.

Afghanistan

The northern part of Afghanistan consists of a large area of elevated ground, over 1000m in height, crossed by many individual mountain ranges that increase in height and coalesce eastwards towards the impressive Hindu Kush Range in the northeastern corner of the country (Figs 7.1, 7.8). Much of the Hindu Kush is over 2000m in height and includes the Kuh-i-Baba Range, at over 5000m, west of Kabul. The highest point in the Hindu Kush is the peak of Tirich Mir (7708m) which is situated across the border in Pakistan. Here, the eastern end of the Hindu Kush Range merges with the Pamir Mountains of Tajikistan, and from there the mountain belt continues south-eastwards as the Karakoram Range. These mountains are the eastern continuation of the northern ranges of Iran described above, and define the northern margin of the Alpine–Himalayan orogenic

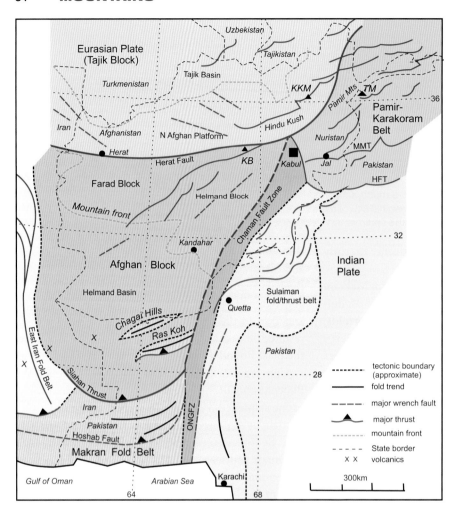

Figure 7.8 Major tectonic features of Afghanistan and adjacent areas. HFT, Himalayan Frontal Thrust; MMT, Main Mantle Thrust; ONGFZ, Ornach–Nal–Ghazaband Fault Zone. Cities: Jal, Jalalabad; Mountains: KB, Kuh-i-Baba; KKM, Kuh-e Khrajeh Mohammad; TM, Tirich Mir. Colours as in Figures 7.2 and 7.5. After Mahmood & Gloaguen, 2012.

belt. North of these mountains is the low-lying desert region of Turkmenistan, which extends west towards the Caspian Sea and northwards into Uzbekistan.

South of the mountains is the flat and relatively featureless region known as the Helmand Basin, which consists mostly of arid desert, and continues southwards into the northern Makran.

The Afghan Block

Most of Afghanistan lies within the Afghan Block – a terrane which, like the Iranian Terrane, is sandwiched between the Zagros–Makran belt in the south and the Eurasian Plate in the north (Fig. 7.5). The Afghan Block is part of the Cimmerian Super-terrane, and consists of Palaeozoic basement of Gondwanan origin overlain by Jurassic and Cretaceous sedimentary cover. The western boundary of the block is defined by the East Iran Fold Belt, which is a complex zone of dextral wrench faulting produced by relative movements between the stable

Iranian and Afghan blocks during the Alpine collision with Arabia. The eastern boundary of the Afghan Block is a major sinistral fault zone that is part of the western transform boundary of the Indian Plate. The northern part of this fault zone is known as the Chaman Fault Zone and the southern as the Ornach–Nal–Ghazaband Fault Zone. East of the fault zone is the complex Sulaiman fold-thrust belt involving the sedimentary cover of the Indian Plate, discussed in the next chapter.

The northern boundary of the Afghan Block is defined by the Herat Fault system, which traverses northern Afghanistan from the Iranian border west of Herat, eastwards to the Hindu Kush mountains north of Kabul, and from there turns north-eastwards into Tajikistan. This zone marks the suture between the Afghan Block and the Tajik Block to the north, which is part of the Eurasian Plate. It resulted from the subduction of Palaeo-Tethys Ocean crust beneath Eurasia during Triassic and Jurassic times, followed by collision at the beginning of the Cretaceous. The Afghan Block itself is an amalgamation of four separate terranes during the early Cretaceous: the Farad Block and the Helmand Block form the main part of the Afghan Block plus two small terranes, the Pamir and Nuristan Blocks, in the northeast corner. The suture zones between these terranes became reactivated during the Alpine compressional event in the Makran Belt and were also affected by the oblique convergence of the Indian Plate. The result of these two processes was the creation of the 700km-wide zone of mountains in eastern Afghanistan, where the northern and southern branches of the Alpine–Himalayan orogenic belt meet at the Chaman Fault Zone (see Fig. 7.8).

The Chagai Fold Belt

There are two small isolated mountain ranges in the southern part of the Afghan Block, near the Afghanistan–Pakistan border, the Chagai and Ras Koh Hills. These are essentially broad anticlines that are related to the Neogene deformation further south in the Makran, but are separated from the remainder of the Makran structures by the broad Hamun-i-Mashkel depression (see Figs 7.3, 7.5).

8

The India–Asia collision zone

As the Alpine–Himalayan mountain belt is followed eastwards from Afghanistan into Pakistan and Tajikistan, there is a marked increase in scale compared to the ranges further west. The belt narrows at the eastern end of the Hindu Kush, as the ranges of the Chaman Fault Zone and the Sulaiman belt merge with it (see Figs 7.1, 7.8). Here the mountain belt is less than 350km across but the central ridge is over 6000m high and culminates in the peak of Tirich Mir, at 7690m. Further east, the belt divides again into a southern branch, consisting of the Karakoram and Himalayan ranges, which curve around the northern fringe of the Indian sub-continent, and a northern branch, which strikes north-eastwards as the Tian Shan, forming the border between Kyrgyzstan to the north and Xinjiang Province of China to the south (Figs 8.1, 8.2). Between the two is the vast expanse of the Tibetan Plateau and the Tarim Basin.

The Karakoram–Himalayan Belt

The Karakoram–Himalayan Belt is usually considered to be the 'type example' of a collisional orogenic belt, but in reality it is unique, as none of the other contemporary orogenic belts are as complete: all the others involve subduction along some or all of their lengths, whereas in the case of the India–Asia collision zone, the continental plates are in contact along the whole length of the belt, all the oceanic plate that formerly intervened having been subducted. The Himalayan sector itself is over 2500km long and is draped around the northern perimeter of the Indian sub-continent, which projects into it at its western and eastern ends. It contains several of the world's highest mountains, including Mount Everest, and is bordered to the north by the high Tibetan Plateau (Fig. 8.3). The western (Pakistan) sector is known as the Karakoram.

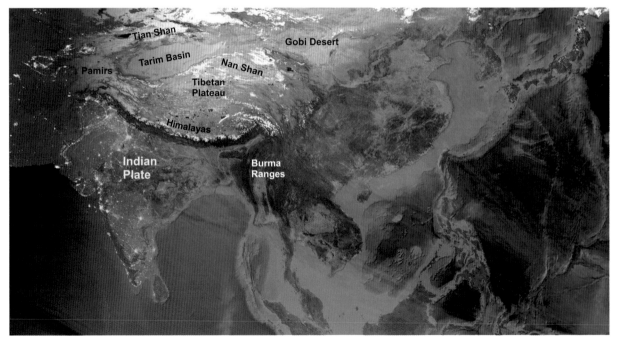

Figure 8.1 Asia from Space. The curved line of the Himalayas is outlined by the green forested zone and is bounded to the north by the Tibetan Plateau. Note the oval shape of the Tarim Basin and the curve of the mountain ranges around the northeast corner of India, extending down through Burma to the Malaysian Peninsula. © Shutterstock, by VanHart.

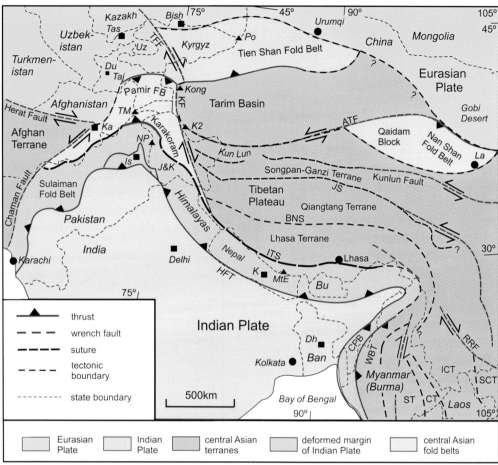

Figure 8.2 Main tectonic features of the Himalayan Belt. Main faults: ATF, Altyn Tagh Fault; KF, Karakoram Fault; RRF, Red River Fault; TFF, Talas–Fergana Fault. Sutures: BNS, Bangong–Nujiang Suture; JS, Jinsha Suture; ITS, Indus–Tsangpo Suture. Tectonic zones: CPB, Chin–Paktai Belt; CT, Chanthaburi Terrane; ICT, Indo-China Terrane; SCT, S. China Terrane; ST, Sibumasu terrane. States, etc: Ban, Bangladesh; J&K, Jammu & Kashmir; Kaz, Kazakhstan; Kyrgyz, Kyrgyzstan; Taj, Tajikistan; Uz, Uzbekistan, Cities: Bish, Bishkek; Dh, Dhaka; K, Kathmandu; Ka, Kabul; La, Lanzhou; Tas, Tashkent. Mountains: Kong, Kongur Shan; MtE, Mt. Everest; NP, Nanga Parbat; Po, Pobeda Peak; TM, Tirich Mir. After Molnar & Tapponnier, 1975; and Searle *et al.*, 2006.

Figure 8.3 The Himalayas. Aerial view looking south from the brown, arid Tibetan Plateau. Mount Everest is in the centre of the view, the black north face prominent. NASA image.

The northern margin of the Indian Plate is shaped like a southerly convex arc, with two prongs at either end, jutting into Asia. The more westerly prong has pushed into northern Pakistan and Tajikistan, forcing the uplift of the mountains of Jammu and Kashmir and the arc of the Pamir Range. The belt here is 750km across from north to south and includes many peaks over 6000m in height, including Nanga Parbat at 8126m. From here, the Karakoram Range extends south-eastwards, at the western end of the Himalayas. The Karakoram Range includes the famous mountain known as K2 which, at 8611m, is the world's second highest peak. Further east, in the Nepal Himalaya, is Everest itself (8848m). The mountain belt here (Fig. 8.3) is narrow, only about 200km wide, and forms the southern fringe of the Tibetan Plateau, which has a mean elevation of over 5000m.

At the eastern end of the Himalayan Range, in the northeast corner of India, the mountain belt turns abruptly through 180°, around the valley of the Brahmaputra River, and divides into several branches, lower and less impressive than the Himalayas, which run roughly southwards through Burma. The pattern of mountain ranges here clearly reflects the shape of the northeastern prong of the Indian Plate.

Plate-tectonic context

The shape of the orogenic belt is dictated largely by the outline of the northern margin of the continental part of the Indian Plate, which is thought of as a relatively rigid 'indenter' moulding the southern margin of the less rigid, more deformable, Asian continent. This model was proposed in 1975 in an influential paper by Peter Molnar and Paul Tapponnier, who interpreted a number of recent or active structures within the Asian continent, many of them located far from the plate boundary, as the result of the collision. The basic idea is reminiscent of the mobilistic model put forward more than sixty years previously by F.B. Taylor and Alfred Wegener (see chapter 2), and departs from the notion, inherent in the original plate-tectonic theory, of the continental crust behaving as an undeformable entity, all parts of which move in the same direction with the same angular velocity.

The southern part of Central Asia with which the Indian continent came in contact is an amalgamation of several continental blocks making up the Tibetan Plateau: the Lhasa, Qiangtang and Songpan-Ganzi Terranes, together with the Tarim Basin to the north,

all of which joined the Siberian core of Asia during the Mesozoic (Fig. 8.2). The sutures between these blocks represent lines of weakness within the Eurasian continent that were exploited during the India–Asian collision, and along which renewed activity in the form of thrusting or strike-slip faulting took place.

Our knowledge of the history of the Himalayan belt is assisted by detailed information from the Indian Ocean magnetic-stripe data (Fig. 8.4), from which it is deduced that India became detached from the Gondwana supercontinent during the early Cretaceous. Around 50Ma ago the convergence rate slowed from between 100 and 180mm/a to nearer 50mm/a, suggesting that the initial contact between the two continents may have

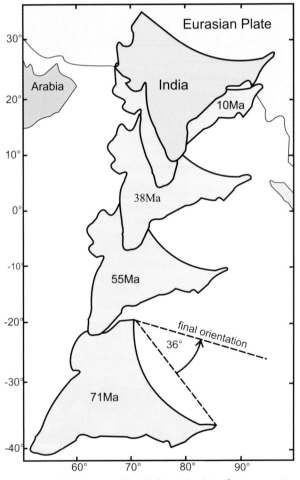

Figure 8.4 Movement of India from the late Cretaceous to the Present. Successive positions based on ocean-floor magnetic anomaly data, with respect to the present position of Asia. Note (1) that the southern margin of Eurasia would have been much further south at the time of initial contact; and (2) that India has experienced a 36° anti-clockwise rotation during its northward journey. After Molnar & Tapponnier, 1975.

occurred then. Prior to this event, subduction of the oceanic part of the Indian plate had been taking place beneath Asia from Cretaceous times. The climax of the collision event, resulting in the uplift of the Himalayan range itself, occurred in the Miocene around 20Ma ago, although uplift and thrusting continue today.

Although the exact shape of the Asian margin prior to the final collision is unknown, it is clear from Figure 8.4 that the first point of contact would have been at the western prong of India. This probably accounts for the anti-clockwise rotation of the sub-continent subsequent to the initial contact, and may at least partly explain the greater degree of deformation and uplift at the western end of the belt. The convergence direction along the northern sector of the belt was approximately at right angles to it, whereas along both sides, it is highly oblique:

the relative movement along both the western and eastern sides of the sub-continent was transpressional, with a large strike-slip component.

The central Himalayan sector

The central sector of the Himalayan belt (Figs 8.2, 8.5) forms an arc, 1750km long and 250km across, extending from northwestern India, through Nepal, to Bhutan, and includes southernmost Tibet. It is bounded in the north by the Indus–Tsangpo Suture, which marks the junction between the Indian and Eurasian plates. In the central sector, the suture is a south-directed thrust that is offset by the north-dipping Renbu-Zedong Thrust. The southern margin of the belt is defined by the Himalayan Frontal Thrust, which marks the edge of the fold-thrust belt on the Indian foreland. The central Himalayan

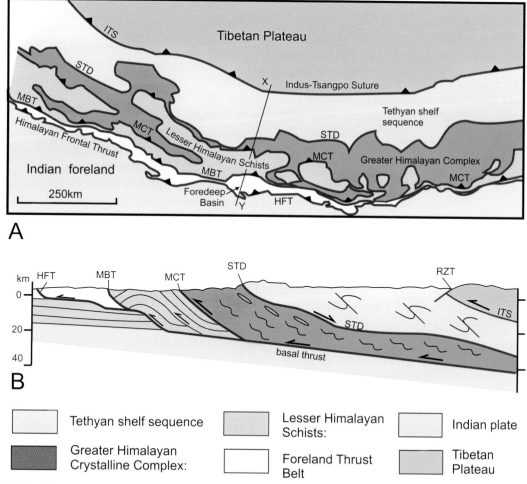

Figure 8.5 The Himalayan belt. Map (**A**) and cross-section X–Y (**B**) of the central (Nepal) sector between the Tibetan Plateau and the Indian foreland showing the main tectonic features. HFT, Himalayan Frontal Thrust; ITS, Indus–Tsangpo Suture; MBT, Main Boundary Thrust; MCT, Main Central Thrust; RZT, Renbu–Zedong Thrust; STD, South Tibetan Detachment. After Harrison, 2006.

belt consists of four separate tectonic zones: these are, traversing from south to north, the Foreland-Thrust Belt, the Lesser Himalayan Schists, the Greater Himalayan Crystalline Complex, and the Tethyan Shelf (Fig. 8.5).

The Foreland Thrust Belt

This unit consists of south-directed fold-thrust sheets involving the Siwalik Formation, which is composed of unmetamorphosed clastic sediments formed in the foredeep basin and derived from the rising Himalayan Mountains.

The Lesser Himalayan Schists

This unit consists of slates and schists derived from clastic sediments of Proterozoic age, originally laid down on the Indian passive margin and deformed into south-directed fold-thrust packages. The southern boundary is marked by the Main Boundary Thrust.

The Greater Himalayan Crystalline Complex

This zone also consists of metamorphosed Proterozoic clastic sediments from the Indian passive margin, but here they have been transformed into high-grade schists and gneisses. The metamorphic grade is inverted, and increases upwards. Near the upper margin is a zone of granite intrusions with a generally lensoid shape. The southern, or lower, boundary of the crystalline complex is formed by the Main Central Thrust, which is actually a ductile shear zone several kilometres wide.

The Tethyan Shelf

This zone consists of largely unmetamorphosed Cambrian to Eocene marine strata, mainly carbonates and shales, originally laid down on the Indian continental shelf. The lower boundary of this zone is a north-dipping normal fault known as the South Tibetan Detachment (STD).

The upper boundary of the Tethyan Shelf, corresponding to the northern margin of the central Himalayan sector, is the south-dipping Indus–Tsangpo Suture zone, which contains an ophiolite complex of Cretaceous to early Cenozoic age. Immediately north of the suture in the central sector is an elongate granite intrusion, the Gangdese batholith (see Fig. 8.9), which represents part of a volcanic arc resulting from the subduction of Indian oceanic lithosphere in the early Cenozoic.

Tectonic history

The three northern tectonic zones of the central Himalayan belt all represent packages of material scraped off the top of the continental Indian Plate as it descended beneath the Asian Plate after the initial collision; these were emplaced as thrust sheets, as indicated in Figure 8.5B. It is thought that the thrust sheets developed by propagating forwards in the manner of other typical thin-skinned foreland fold-thrust belts already described. As each sheet ramped up, the orogen would shorten and thicken, although erosion would continuously remove material from the roof of the structure.

Two alternative explanations have been put forward to explain the juxtaposition of the high-grade gneisses of the Greater Himalayan Complex and the Tethyan Shelf sediments above them, which are separated by the South Tibetan Detachment. It was formerly thought that successive forward-propagating thrusts would cause the Greater Himalayan Complex and its Tethyan sedimentary cover to be arched up and subjected to erosion. Gravitational spreading then caused the Tethyan cover to slide down to the north, exposing the higher-grade rocks of the crystalline complex. A more recent explanation relies on a mechanism termed 'channel-flow'. In this model, the metamorphic rocks of the Greater Himalayan Complex are regarded as a hot, partially molten piece of Indian crust that has flowed under gravitational pressure, from its original position beneath the Asian plate, upwards to the surface between the Tethyan Shelf unit and the Lesser Himalayan Schists beneath. It is possible that both mechanisms may have contributed to the present structure.

The Karakoram and the Pamirs
The Karakoram

The Karakoram Range is a relatively short, arcuate mountain belt at the extreme north-westernmost corner of the Indian Plate and includes several very high peaks, including Nanga Parbat (8126m). In geological terms, the southern part of the range lies within the Himalayan fold/thrust belt, situated on the Indian Plate, between the Himalayan Frontal Thrust and the Indus–Tsangpo Suture (or Main Mantle Thrust), which here is a wide, north-dipping shear zone forming the southern boundary of a volcanic arc terrane known as the Kohistan Arc (Fig. 8.6). Kohistan is the mountainous district in the northwestern corner of Pakistan, bordering Kashmir to the east and Afghanistan to the west and north.

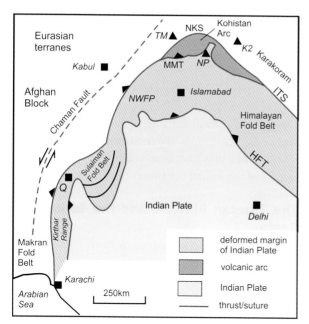

Figure 8.6 The Kohistan Arc and the Sulaiman Fold Belt. HFT, Himalayan Frontal thrust; ITS, Indus–Tsangpo Suture; MMT, Main Mantle Thrust (=ITS); NKS, North Karakoram Suture. NWFP, Northwest Frontier Province. Mountains: K2, (Godwin Austen); NP, Nanga Parbat; TM, Tirich Mir. Cities: Q, Quetta. After Butler & Prior, 1988.

The Kohistan arc terrane is only exposed over a length of about 500km at the northwestern apex of the belt and has not been traced further to the east or west. The arc itself is bounded on its north side by the Northern Kohistan Suture, which marks the southern boundary of the Asian Plate. The Kohistan Arc is believed to have originated as a volcanic island arc, of late Jurassic to early Cretaceous age, which collided with the Asian continent in the mid-Cretaceous (Fig. 8.7). The subsequent collision with the Indian Plate around 50Ma ago, and the subsequent anti-clockwise rotation of India, resulted in a concentration of compressional stress and rapid uplift, both of the Kohistan Arc and of the basement of the Indian Plate, of about 10km over the last 10Ma – a much faster rate than that of the rest of the Himalayan Belt.

The Pamirs

The region known as the Pamirs is a high mountainous plateau crossed by several separate mountain ranges. The mean elevation is even higher than the adjacent Tibetan Plateau, and there are many peaks over 6000m high with large permanent ice-fields. The highest peak is Kongur, at 7649m, near the eastern margin of the

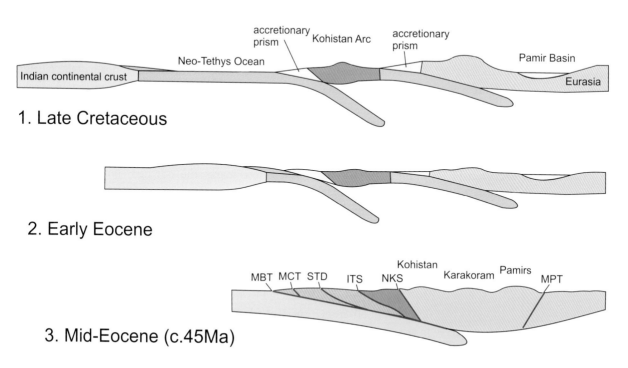

Figure 8.7 Cartoon sections across the Kohistan–Pamir region. Possible arrangement of the main tectonic crustal units from the late Cretaceous to the final collision in the Eocene. ITS, Indus–Tsangpo Suture; MBT, Main Boundary Thrust; MCT, Main Central Thrust; MKT, Main Karakoram Thrust; MPT, Main Pamir Thrust; NKS, North Kohistan Suture; STD, South Tibetan Detachment.

plateau, just across the Chinese border (see Fig. 8.2). The Pamir region lies mostly in Tajikistan and eastern Afghanistan, and is situated at the confluence of the northern and southern branches of the Alpine–Himalayan Belt. The Hindu Kush and Chaman Fault Zone join it from the west (see Fig. 7.8), and on its eastern side, the Tian Shan extends north-eastwards and the Karakoram south-eastwards, around the north and west sides, respectively, of the Tibetan Plateau.

The Pamir region is situated entirely within the Eurasian Plate, but has experienced the effects of the Himalayan collision event, in terms partly of thrust-related deformation and partly of uplift. On its southern side it is bounded by the arcuate, north-dipping, North Kohistan Suture, and in the north by a similarly arc-shaped south-dipping thrust – the Main Pamir Thrust (Fig. 8.7). A second arcuate south-dipping thrust occurs in the middle of the block. The effect of the Himalayan compression was to elevate the whole block, exposing Eurasian basement in the southern sector. Overlying this basement is a sedimentary platform cover ranging from Cambrian to Recent in age.

The Sulaiman Fold Belt

The western side of the Indian Plate is bounded by the Ornach Nal-Ghazaband-Chaman strike-slip fault system (see Fig. 7.8). East of these faults, which effectively constitute a transform fault zone, is a fold-thrust belt that lies entirely within the Indian Plate, and involves the sedimentary cover of the Indian platform. This belt follows two separate mountain ranges: the more southerly, the Kirthar Range, extends for 550km through Baluchistan from Karachi in the south to the city of Quetta; the more northerly forms a great southeast-facing arc, known as the Sulaiman Mountains, which run for about 400km from Quetta northwards to the Northwest Frontier Province, where the fold belt turns eastwards (Fig. 8.6). The mountain ranges are generally over 2000m but decrease in height southwards; the highest point is Takht-e Sulaiman (Throne of Solomon) at 3487m, just east of Quetta.

The Sulaiman Fold Belt is interesting in that the arcuate mountain ridges curve through about 60° from E–W in the south to NNE–SSW in the north, following the trend of the major folds, so that only in the southern part of the range are the folds at right angles to the India–Asia convergence direction (i.e. NNW–SSE); those in the northern sector are highly oblique. However, data from recent seismic activity indicate that the active deformation was accomplished at least partly by movements on south-dipping reverse faults, which must therefore not be directly connected to the surface features in the northern sector of the belt.

The marked bend in the fold belt at Quetta coincides with a bend in the Indian Plate boundary, which at this point diverges from the line of the Chaman Fault to strike north-eastwards, suggesting that the shape of the Sulaiman belt may have been controlled by the original margin of the Indian continent.

The Tibetan Plateau and the Tarim Basin

The vast Tibetan Plateau extends for over 2400km from the edge of the Pamir Mountains in the west to near the city of Lanzhou in Gansu Province in the east. It is 1300km across at its widest, but narrows to the west. It is everywhere over 4000m in height and includes several impressive mountain ranges containing peaks over 7000m high. The southern margin of the plateau is defined by the Himalayan and Karakoram Ranges, while the northern boundary is the Altyn Tagh Fault, which separates the plateau from the Tarim Basin (see Fig. 8.1). The northeastern boundary is defined by the Qilian Shan (or Nan Shan) Range at the southern edge of the Gobi Desert. There are also several less prominent ranges within the Plateau; these correspond to re-activated faults or sutures.

The Tarim Basin lies immediately north of the Tibetan Plateau in the Chinese Province of Xinjiang. In abrupt contrast to the plateau, the mean elevation of the basin is below 1000m, although it rises in height towards the west. The basin has a roughly oval shape and is bounded in the north by the Tian Shan (or Tien Shan) Range, which here marks the northern limit of Himalayan compressive deformation.

Tectonic summary

The geology of the region north of the Indus–Tsangpo suture is less well known than that of the Himalayan belt. This part of the Eurasian plate has been less obviously affected by the Himalayan compressional deformation, part of which is concentrated along the suture zones between the several separate terranes or microplates that had previously accreted to Central Asia and form part of the large Cimmerian 'super-terrane' discussed in the previous chapters. These are, from south to north, the Lhasa, Qiangtang, Songpan–Ganzi and Tarim Terranes, together with the Qaidam block in the northeast.

The various terranes and crustal blocks originated by splitting off from Gondwana at different times and travelled separately to join Eurasia: Tarim and Qaidam left in the Devonian and joined Eurasia in the early Permian; the Qiangtang (part of Cimmeria) left in the Permian and joined in the Triassic; and the Lhasa block left in the Triassic and joined in the early Jurassic. The Songpan–Ganzi Terrane is an accretionary complex sandwiched between these continental blocks and the Qiangtang Terrane.

Himalayan deformation has also been accommodated by movements along a network of strike-slip (wrench) faults. These faults form a conjugate set: a sinistral set varying in orientation from NE–SW to ENE–WSW, and a dextral set varying from N–S to WSW–ESE. Towards the east, the faults of both sets curve round into a more N–S orientation as the belt of deformation turns southwards into Burma and Indo-China (see Fig. 8.1). Numerous north–south oriented graben systems also indicate significant E–W extension and this, coupled with the movements on the numerous minor conjugate wrench faults, has been interpreted to indicate that much of the north–south convergence between India and Asia has been accommodated by the sideways extrusion of Asian crust.

Repeated precise GPS measurements have enabled accurate movement vectors across the orogen to be calculated (Fig. 8.8); these range generally between 5 and 15mm/a, confirm the lateral extrusion model, and also indicate a gradual diminution of flow velocity

northwards; there is thus no clearly defined northern margin to the deformation resulting from the collision, as there is in the south. Data on slip rates derived from GPS and satellite radar measurements along the wrench faults are typically in the range 5–15mm/a. Interestingly, the slip rates on the very large wrench faults such as the Altyn Tagh and Karakoram Faults, formerly believed to have accommodated the bulk of the north–south compression, are little different from the others.

This pattern of distributed deformation, achieved mainly by faulting, applies only to the upper (seismogenic) crust. Beneath this, the middle crust, being warmer and more ductile (see below), will have deformed in a more continuous manner, employing shear zones rather than discrete faults.

The origin of the Tibetan Plateau

There has been a vigorous debate that has lasted for many decades about the reason for the high elevation of the plateau. Orthodox plate-tectonic theory originally suggested that the explanation was the under-thrusting of the Indian Plate, causing a doubling up of the Tibetan crust. However, this would imply a very shallow-dipping subducting slab extending for over 1000km, which mechanically seemed unlikely. Seismic data show that the present base of the Indian crust descends from about 40km depth at the orogenic front to 70km beneath the Indus–Tsangpo Suture and remains around that level for about 200km across South Tibet before it disappears, to be replaced by the quite different structure of the Tibetan Plateau, as described below. As the subducting slab now extends for only a relatively short distance beneath the southern part of the Plateau, as indicated in Figure 8.9, it cannot be responsible for the plateau uplift.

More recent intensive research employing various indirect geophysical techniques has revealed that the crust of the Tibetan Plateau is considerably warmer, and therefore much weaker, than the Indian equivalent, and that its thickness is over double that of the Indian, varying from 70km in the south to around 60km in the north, with a marked change across the Bangong–Nujiang Suture (Fig. 8.9). Moreover, as a result of the different physical properties in the lower part of the crust, the brittle, fault-dominated tectonics of the upper crust appears to be replaced in the lower crust by a pattern of general eastwards flow.

The behaviour of the lithospheric mantle beneath Tibet is more difficult to determine. Because of the marked increase in strength between the lower crustal

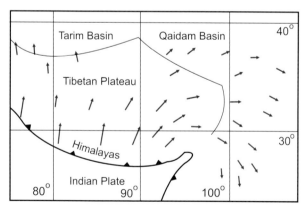

Figure 8.8 Active deformation of the southern part of the Eurasian Plate. Velocity vectors (red arrows) showing direction and relative amount of movement in the deformed part of the Eurasian Plate relative to stable Eurasia. The pattern shows that the flow is to the north, gradually decreasing northwards in the western sector, while directed north-eastwards, then eastwards and ultimately south-eastwards in the eastern sector. After Searle *et al.*, 2011.

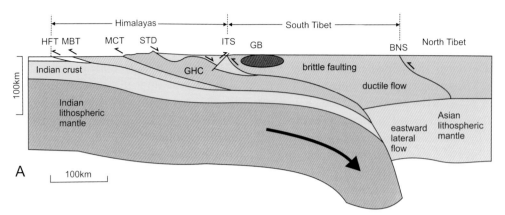

Figure 8.9 Simplified cross-section through the central sector of the Himalayan Belt. Note contrasting arrangement of crust and mantle lithosphere in the Indian and Eurasian Plates. GB, Gangdese Batholith; other abbreviations as in Figure 8.5. Note that the Asian crust is much thicker than the Indian, and that a considerable amount of Asian lithospheric mantle is apparently missing. After Searle *et al.*, 2011.

and uppermost mantle material, the latter would not be expected to behave in the same ductile fashion. Nevertheless, the seismic properties of the Tibetan mantle contrast with those of the Indian lithospheric mantle and are consistent with a model of mantle flow similar to that inferred for the crust, at least for the material below the topmost part. The 'missing' Tibetan mantle lithosphere caused by the under-thrusting of the Indian lithosphere, evident from Figure 8.9, could thus be explained in the same way as the Tibetan lower crust, with the exception that the lower part may have been absorbed into the asthenosphere, leaving the total thickness more or less unchanged.

The conclusion drawn from these observations is that much of the compressive stress generated by the Indian collision has been accommodated by thickening and shortening of both the crust and lithospheric mantle of the Tibetan Plateau, accompanied by sideways transfer of material, mostly to the east and southeast. The height of the plateau can be explained by its warmth and thickness, and its relative flatness by the fact that general isostatic equilibrium can be achieved by the ease of flow in the lower crust. The high heat flow is attributed to large quantities of igneous material within the crust, mostly arising from the pre-collision subduction process.

In contrast to the Tibetan example just described, the Indian lithosphere is believed from seismic evidence to be relatively cool and strong, providing a broad, semi-rigid slab which has underthrust the Asian plate for a distance of over 200km north of the Indus–Tsangpo Suture. The northward movement of this slab is the main driver for the deformation of the orogen. The current convergence

rate between the Indian and Eurasian Plates is about 50mm/a towards the north. Of this, approximately half is taken up by shortening across the Himalayan fold-thrust belt and half by the shortening and eastwards extrusion of the Tibetan lithosphere – perhaps 80% by shortening and thickening, and 20% by lateral extrusion. Much of the shortening across the Himalayas is accomplished by crustal thickening, a considerable proportion of which has been removed by erosion.

The northern mountain belts

The northern branch of the Alpine–Himalayan mountain belt contains two separate mountain ranges, the Tian Shan and the Qilian Shan, or Nan Shan, which together extend from the edge of the Pamir Plateau in the west to near the City of Lanzhou in the east – a combined distance of around 2600km. The Tian Shan–Nan Shan line defines the northern limit of the main Himalayan deformation, but movements also occurred north of the Nan Shan in the Altai Range of southwest Mongolia (see chapter 14 and Fig. 14.13), where some thrusts and dextral strike-slip faults remain active.

The Tian Shan

This mountain range, alternatively spelled 'Tien Shan', stretches for 1800km in a great arc from the eastern edge of the Pamir Plateau towards the Mongolian border, forming the northern rim of the Tarim Basin (see Fig. 8.2). It is a substantial mountain range, rivalling the Hindu Kush and the Pamirs in scale: the main ridge in the west lies along the boundary between China and Kyrgyzstan, and is everywhere well over 4000m in height, reaching 7439m at the summit of its highest peak, Pik Pobedy (Jangish Choksu).

The Tian Shan is an example of an intra-plate orogenic belt – the Himalayan deformation exploited weak zones within the Eurasian Plate that had resulted from two late Palaeozoic subduction–collision episodes: the older created a north-dipping suture along the southern margin of the belt, and the younger a south-dipping suture on the opposite side of the belt. The Himalayan collision re-activated the former suture zones into steep thrusts that elevated the wedge-shaped block between to form the mountain belt. Most of the deformation is concentrated along the boundary fault zones and the interior of the block experienced minimal effects. In addition to the thrusts, which are the predominant expression of the Himalayan contraction, there are also several dextral strike-slip faults that cut the thrusts – the most important being the NW–SE Talas–Fergana Fault, which forms the western boundary of the Tian Shan. The belt is strongly active seismically, having experienced many recent earthquakes: it has been estimated that the convergence rate across the belt is around 20mm/a, mainly over the last 10Ma, and has resulted in up to 20km of crustal shortening parallel to the convergence direction across the Himalayas.

The Nan Shan (or Qilian Shan)

This range extends for 800km along the northeast side of the Qaidam Block (Fig. 8.2), ending near Lanzhou in the valley of the Yellow River (Huang He). The latter two-thirds of the range follows the south side of the Great Wall. This range is smaller in scale than the Tian Shan: although it contains several peaks over 6,000m in height, it is more accurately described as the northern margin of the high Tibetan Plateau, from which it differs only slightly in elevation. However, the southwest-dipping thrust along the northeastern margin of the Nan Shan is strongly active: a magnitude 7.6 earthquake in 1927 is thought to have been responsible for nearly 41,000 deaths, as well as widespread structural damage.

The Eastern Sector

East of Tibet, the Alpine–Himalayan mountain belt divides into two main branches as it turns southwards through Burma (Myanmar) into Indochina (Fig. 8.2). The more westerly, relatively minor, branch follows the eastern margin of the Indian Plate forming the Patkai, Chaga and Chin ranges. The more easterly branch descends southwards through eastern Burma and southwestern China as a 1000km-wide belt containing many roughly parallel mountain ranges. In the northern sector of this belt, the individual mountain ranges curve around the northeastern corner of the Indian Plate from the heights of the Tibetan Plateau, separated by the valleys of the Irrawaddy and Salween rivers and the upper reaches of the Yangtze. One of these ranges, the Daxue Shan, contains the impressive peak of Gongga Shan, at 7556m the highest peak east of the Himalayas.

The Chin–Naga–Patkai Fold Belt

This belt of mountains extends for about 950km from Cape Negrais in southwest Burma (see Fig. 9.2) to the northeastern corner of India, where it meets the Himalayas. The belt, which is up to 150km wide, includes four separate ranges, from south to north: the Arakhan, Chin, Naga and Patkai Hills. The northern part of this mountain belt, from the Chin to the Patkai Ranges, forms the eastern boundary of the Indian Plate, and consists of a number of separate parallel ranges which together form an arc facing west towards the interior of the Indian Plate. The highest point in the belt is Saramati, at 3826m, in the Naga Range. The southern extension of the belt, known as the Arakhan, or Rakhine Mountains, lies wholly within Burma and is discussed in the following chapter.

The origin of the fold belt lies in the anti-clockwise rotation of the Indian Plate after its initial contact with Eurasia. This rotation took place about a hinge at the northwest corner of the sub-continent and resulted in eastwards convergence across the western border of Burma. The structure is that of a typical thin-skinned fold-thrust belt involving a Palaeocene to Recent platform cover more than 6km thick overlying Indian Pre-cambrian basement (Fig. 8.10). The belt is made up of an outer zone of thrust slices, the Schuppen Zone, bounded on its western side by the Naga Thrust. This boundary thrust separates the fold belt from the un-deformed Assam Shelf, in the northeastern part of the Indian Plate. Several other major thrusts occur east of the Schuppen Zone, bringing successively deeper parts of the Cenozoic sedimentary sequence to the surface.

On the eastern side of the fold-thrust belt lies the Central Burma Basin, which occupies the southern extension of the Lhasa Terrane, squeezed eastwards and southwards by the effects of the Himalayan collision (see Fig. 8.2).

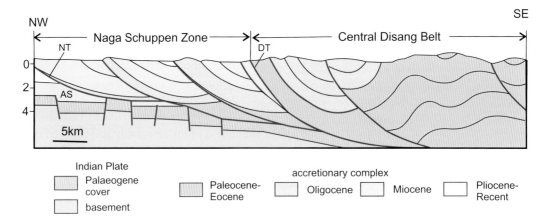

Figure 8.10 The Naga fold-thrust belt of NE India: simplified section. AS, Assam Shelf; DT, Disang Thrust; NT, Naga Thrust. After Oil and Natural Gas Corporation, India, via Wikimedia Commons.

East of the Central Burma Basin is the wide mountain belt of Eastern Burma and the Sichuan Province of SW China. This north–south belt crosses a further four distinct tectonic zones. From west to east, these are known locally as: the Sibumasu Terrane (the equivalent of the Qiangtang of Tibet), the Chanthaburi Arc Terrane (the Songpan–Ganzi of Tibet), the Indochina Terrane and the South China Terrane. These zones are described in the following chapter. The major sutures separating these terranes each experienced renewed compressional effects as a result of the Himalayan convergence.

Tectonic summary

The effect of the Indian collision during Eocene to Oligocene time was to force the various Gondwanan terranes of Tibet eastwards and south-eastwards, wrapping them around the northeastern corner of the Indian Plate, and squeezing them against the more stable blocks to their east (see Figs 8.2, 8.8). The large mountainous region that embraces eastern Burma, northern Laos and southwest China is, at least partly, the result of compression and uplift responding to the north-eastwards motion of the continental part of the Indian Plate.

9

Southeast Asia

After turning southwards through Burma and southwest China, the Alpine–Himalayan mountain belt follows a north–south trend into Thailand and the Malaysian Peninsula, then curves round eastwards in a great arc into Western Indonesia (Fig. 9.1). The more westerly, relatively minor, branch follows the eastern margin of the Indian Plate as the Arakhan Range, the southern extension of the Patkai–Chaga–Chin belt discussed in the previous chapter. The Arakhan belt continues as a subduction zone along the coast of Burma to Cape Negrais, forming the eastern margin of the oceanic part of the Indian Plate. From this point, the western margin of the mountain belt is an island arc that curves in a wide sweep through the Andaman and Nicobar Island chains and continues through Sumatra and Java as the Indonesian Island Arc.

The more easterly branch of the Alpine–Himalayan mountain system descends southwards through eastern Burma and southwestern China as a 1000km-wide belt containing many roughly parallel mountain ranges. These ranges decrease in height southwards until they gradually die out in northern Thailand and Indochina.

Most of northern Thailand is a featureless plain, except for a narrow mountain chain that runs down the centre of

the country to just northeast of Bangkok, following the eastern margin of the Chanthaburi Terrane (see below). A second narrow mountain range, rising to over 2000m in southern Burma, follows the Burma–Thailand border and continues down the spine of the Malaysian Peninsula, where several peaks in the Cameron Highlands are over 2000m. The latter range, which is over 2000km long, lies within the Sibumasu Terrane, discussed below. Between the western and eastern branches of the mountain system lies the wide Andaman Sea, which is an active back-arc basin 1000km in length and 600km wide.

The Southeast Asian Fold Belt
Tectonic context

Southeast Asia is composed of a mosaic of separate continental terranes that have Gondwanan origins, and have joined the Siberian core of Eurasia at different times. The largest of these are the North China, South China and Indochina–East Malaya blocks, which are largely stable areas founded on Precambrian basement. All these stable blocks left Gondwana in the Devonian Period but travelled separately towards Siberia: Indochina joined South China during the early Carboniferous and the

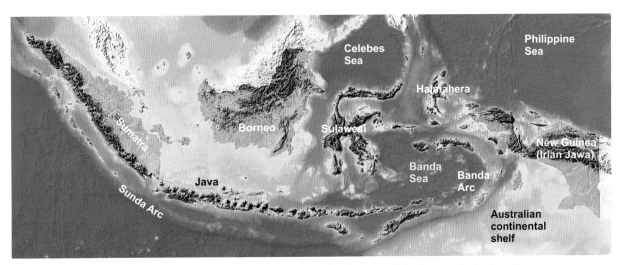

Figure 9.1 Relief map of Indonesia showing the main topographical features. © Shutterstock, by Arid Ocean.

combined South China–Indochina–East Malaya block joined North China in the late Triassic. At about the same time, the Sibumasu Terrane, which represents the southernmost extremity of Cimmeria, also became attached to the combined South China–Indochina–East Malaya 'super-terrane', which finally joined Eurasia during the Jurassic.

The effect of the Indian collision during Eocene to Oligocene time was to force the Gondwanan terranes of Tibet eastwards and south-eastwards, wrapping them around the northeastern corner of the Indian Plate, and squeezing them against the more stable blocks to their east (see Figs 8.2 and 8.8 in the previous chapter). The large mountainous region that embraces eastern Burma, northern Laos and southwest China is, at least partly, the result of compression and uplift responding to the north-eastwards motion of the continental part of the Indian Plate. The southeast Asian sector of the Alpine–Himalayan orogenic belt thus falls into two distinct parts: the northern, running through northern Burma and Southwest China, resulted from the convergence between the Indian and Eurasian Plates, whereas the southern part represents the response to the convergence between the Australian Plate and Eurasia, and comprises southern Burma, Thailand and Indochina, together with the Malay Peninsula and the island arc system extending from the Andaman Islands to Indonesia.

Tectonic zones of the Southeast Asian Fold Belt

This north–south belt can be divided into six distinct tectonic zones east of the Indian and Australian plate boundaries (Fig. 9.2). These are: from west to east, the Western Accretionary Zone, the West Burma Terrane, the Sibumasu Terrane, the Chanthaburi Arc Terrane, the Indochina Terrane and the South China Terrane. These terranes are separated by major sutures, formed at different times, each of which has experienced renewed compressional effects as a result of the Himalayan convergence.

The fold belt is flanked on its western side by a subduction zone running parallel to the eastern coast of the Bay of Bengal. Along this section of the plate boundary, the oceanic part of the Indian Plate is moving north-eastwards beneath Asia. South of Cape Negrais, the western margin of the fold belt is replaced by the Andaman–Nicobar island arc chain, which follows the subduction zone marking the boundary of the oceanic part of the Australian Plate. The two oceanic plates are separated by a N–S-trending transform fault – the Ninety-East Ridge (see Fig. 10.1); east of this fault, the northeasterly convergence direction is unchanged, although the convergence rate is faster. The two plates are often regarded as a single Indo-Australian Plate, but the difference in convergence rates between them has produced a zone of deformation west of the Nicobar Islands that will eventually result in a new plate boundary.

The Western Accretionary Zone

This zone corresponds to the Chin–Naga–Patkai fold-thrust belt discussed in the previous chapter, and consists of the deformed Cenozoic sedimentary cover of the Indian Plate (see Figs 8.2, 8.10). Its western boundary is the Naga Thrust, which defines the western margin of the fold belt, and the zone is bounded in the east by the Mawgyi Suture, which separates it from the West Burma block. The Mawgyi Suture is the most westerly and most recent of the sutures separating the six terranes, having finally closed during the Himalayan collision in the Oligocene. It corresponds to the Indus–Tsangpo suture in Tibet.

South of the Bangladeshi-Burma border, the onshore part of the zone is represented by the Arakhan mountain range, but the greater part of the zone lies offshore between the subduction zone and the mainland coast. South of Cape Negrais, the entire zone swings offshore to become the Andaman–Nicobar Island Arc, which is separated from the Burmese mainland by the active Andaman Sea back-arc basin, formed 4Ma ago. The island chain consists mainly of Neogene sedimentary rocks forming an accretionary complex similar to that of the Arakhan Range to the north, and is strongly active seismically, having experienced several severe earthquakes in recent years, including the disastrous Banda Aceh event in 2004. The arc curves round to a NW–SE orientation through Sumatra, where part of the zone appears onshore. Here, the zone is bounded on its northeast side by the West Sumatra Block, which is regarded as the probable continuation of the West Burma Block, with the same origin. The northern part of the arc is highly oblique to the convergence direction of the Australian Plate and is cut by numerous N–S sinistral strike-slip faults. Because present-day motion in this sector makes such a small angle to the plate boundary, there is no active vulcanicity.

The West Burma Terrane

This relatively stable block, centred on the Irrawaddi Valley, separates the Arakhan mountain range in the

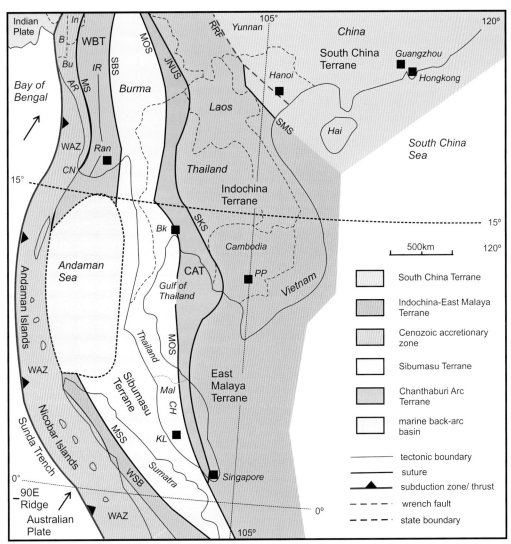

Figure 9.2 Main tectonic units of SE Asia. Note: the Indonesian islands east and southeast of Malaysia are shown in Figure 9.6. CAT, Chanthaburi Arc Terrane; JNUS, Jinghong–Nan Uttaradit Suture; MOS, Main Ocean Suture; MS, Mawgyi Suture; MSS, Mid-Sumatra Suture; RRF, Red River Fault; SBS, Shan Boundary Suture; SKS, Sra Kaoe Suture; SMS, Song Ma Suture; WAZ, Western Accretionary Zone; WBT, West Burma Terrane; WSB, West Sumatra Block. Geographical names: AR, Arakhan Range; B, Bangladesh; Bk, Bangkok; Bu, Burma; CH, Cameron Highlands; CN, Cape Negrais; In, India; IR, Irrawaddi River; KL, Kuala Lumpur; Mal, Malaysia; PP, Phnom Penh; Ran, Rangoon. After Metcalfe, 2011.

Western Accretionary Zone from the much higher ranges of eastern Burma and Yunnan Province in China. The terrane is thought to have separated from Gondwana in the Devonian, like the larger Indochina and South China terranes. On its eastern side, it is bounded by the Shan Boundary Suture, separating it from the Sibumasu Terrane. The Shan Boundary Suture is a continuation of the Banggong–Nujiang Suture in Tibet.

South of the Andaman Sea, this suture reappears as the Mid-Sumatra Suture, separating the West Sumatra Block from the Sibumasu Terrane. The Shan Boundary–Mid-Sumatra Suture is thought to be the site of a branch of the Palaeo-Tethys Ocean; however, the present boundary is a strike-slip shear zone and there is no direct evidence of oceanic material.

The Sibumasu Terrane

The name of this terrane is an amalgam of Sino-Burma-Sumatra, reflecting the three regions through which it passes. It is the southernmost part of the long, thin Cimmerian 'super-terrane', represented in Tibet by the Qiangtang Terrane (see Fig. 8.2), which was bent southwards as a result of the Himalayan collision. In common with the other parts of Cimmeria, Sibumasu was detached from Gondwana in the Permian, and joined the Indochina Terrane during the Triassic. Between the two is the narrow volcanic arc terrane known as the Chanthaburi Terrane.

The Sibumasu Terrane is represented in the Shan Ranges of Northern Burma and by the narrow mountain belt that runs down the spine of the Malay Peninsula through the Cameron Highlands to Kuala Lumpur.

The Chanthaburi Arc Terrane

This volcanic arc terrane between the Sibumasu and Indochina terranes extends from Eastern Burma to southern Malaysia, with a large gap where it is inferred to cross the Gulf of Thailand. The suture between it and the Sibumasu Terrane, known as the Main Ocean Suture, is regarded as the site of the former Palaeo-Tethys Ocean, closed during the Triassic Period, and is represented by an accretionary complex. The main part of the terrane consists of a volcanic arc, which is separated from the western margin of the South China–Indochina block by a second suture zone. This second zone is interpreted as a former subduction zone marking the site of a closed back-arc basin, which had opened during the late Carboniferous to early Permian between the volcanic arc and the main part of the Indochina Block, and closed

again in the late Permian. This eastern suture, confusingly, is known by three different names: Jinghong in China, and Nan Uttaradit and Sra Kaoe in Thailand.

The Indochina Terrane

This large, stable block is effectively the foreland of this sector of the orogenic belt. It occupies almost the whole of Indochina and is considered to correlate with the East Malaya Block south of the Gulf of Thailand. In common with South China, it consists of Precambrian basement of Gondwanan affinity and separated from Gondwana in the Permian. It joined South China in the late Triassic and the combined terranes finally joined Asia in the Jurassic.

The South China Terrane

This terrane forms part of the stable foreland of the South Asian fold belt. The suture separating it from the Indochina block is known as the Song Ma Suture in Indochina (equivalent to the Ailaoshan suture in the Chinese sector). It is displaced by the major strike-slip Red River Fault, which has caused the Indochina block to be displaced sinistrally, squeezed southwards by the Himalayan collision.

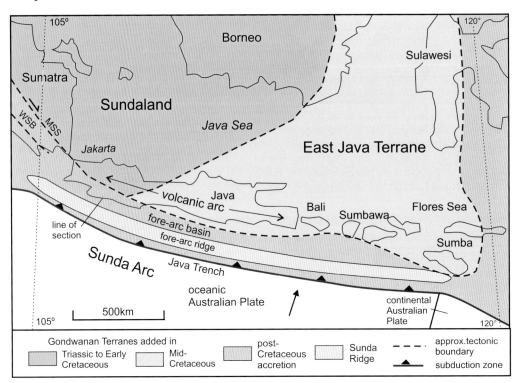

Figure 9.3 The Sunda Arc. Main tectonic units of the Sunda Arc and its foreland. MSS, Mid-Sumatra Suture; WSB, West Sumatra Block. There are a large number of volcanoes on the Sunda Arc, too numerous to show on this map. After Hall, 2011.

Figure 9.4 Mount Sinabung, NW Sumatra. This 2460m peak is typical of the active andesitic volcanoes of the Sunda Arc. Shutterstock©TJUKTJUK.

The Sunda Arc

The Sunda island arc system includes the large islands of Sumatra and Java, together with three smaller islands to their east, ending in the island of Sumbawa (Figs. 9.1, 9.3). The arc system is over 3200km in length – Sumatra alone is about 1750km long and up to 400km across. Both main islands are dominated by impressive mountain ranges containing many active volcanoes; some of these, especially in Java, are well over 3000m in height. The highest is Kerinci, at 3805m, in Sumatra. One of the most active volcanoes is Sinabung, in northern Sumatra (Fig. 9.4) which is situated at the northwest end of an enormous, 100km-long, crater lake. Sinabung has experienced a series of explosive eruptions in recent years, resulting in numerous casualties. Another well-known volcano is the slightly smaller Sibayak, which although no longer actively erupting, exhibits geothermal activity in the form of steam vents and hot sulphurous springs. The Sunda Arc is typical of subduction-related volcanic arcs in exhibiting mainly andesitic vulcanicity and explosive eruptions.

The mountain ranges are narrow, less than 100km wide; in the case of Sumatra they closely follow the southern coast, while the remainder of the island is a broad, featureless plain (Fig. 9.1). The large islands form a wide arc that trends NW–SE in Sumatra and E–W at its eastern end in Sumbawa. About 150km offshore is a second, outer, arc composed of a chain of smaller islands situated on a submerged ridge extending for the whole length of Sumatra but much subdued off Java. The Sunda subduction zone occupies a trench that lies 200–300km offshore from the main islands. This trench, known as the Sunda Trench off Sumatra, and the Java Trench further east, reaches over 6km in depth south of Java but disappears as an obvious topographic feature in the northwestern part of the outer arc.

Tectonic summary

Although currently part of the same active volcanic arc system, the islands of Sumatra and Java differ significantly in their geological composition and history. As shown in Figure 9.2, Sumatra is divided into three tectonic zones: the northeastern part of the island is part of the Sibumasu Terrane described above, which occupies the low-lying plain and is separated from the West Sumatra Block by the Median Sumatra Suture, which is a dextral strike-slip fault. The volcanic arc is situated within the West Sumatra Block. The outermost zone is the accretionary terrane, containing post-Cretaceous sediments deformed in a fold-thrust belt. This outer zone is exposed in the outer coastal parts of Sumatra and in the outer islands, but much of it is submerged.

In Java (Fig. 9.3), the northern foreland is more complex: the western part, known as Sundaland, includes western Borneo and the Java Sea between it and West Java, and is underlain by Gondwana-derived crustal blocks added to East Malaya prior to the mid-Cretaceous. The eastern part, known as the East Java Terrane, is mostly occupied by mid-Cretaceous accretionary material. The island itself, together with the smaller islands to the east, consists almost entirely of volcanoes and volcanic material belonging to the active Sunda Arc, which is situated along the margin of the Sundaland and East Java terranes. The offshore part of the arc differs from Sumatra in that the accretionary prism occupies a submerged ridge, the fore-arc ridge, separated from the islands by the fore-arc basin, which is a 50km-wide elongate trough. Figure 9.5 is a section across the offshore arc structure obtained from refraction seismic data. The trench, which here is about 5km deep, lies more than 50km south of the ridge. The seismic section, one of a series taken across the arc, indicates that the oceanic slab is descending at a relatively shallow angle beneath Java and that the accretionary prism here is over 20km thick. The current convergence rate across the arc is up to 70mm/a in a NNE–SSW direction.

Eastern Indonesia

The part of Indonesia north and east of Java is a confusing jumble of islands of various sizes, the largest being Borneo in the west and New Guinea in the east (Figs. 9.1, 9.6). The Sunda Island Arc is followed eastwards by a string of smaller islands known as the Lesser Sunda Islands. These in turn give way eastwards to the Banda Arc system, which consists of two parallel arcs, which both curve through almost 180° around the Banda Sea. The inner, volcanic arc consists of a series of very small islands linked by a submerged ridge, which ends at the larger island of Seram. The outer, inactive, arc is situated on the margin of the Australian continental shelf and includes the large island of Timor as well as several smaller islands, and ends near the southern coast of New Guinea. All the larger islands are mountainous: the highest peak in the Banda Arc is Binaiya, 3019m, on the island of Seram.

North of the Banda Sea, between Borneo and Seram, is the strangely shaped island of Sulawesi (formerly Celebes). Sulawesi has the appearance of a strange beast whose long neck is twisted round to face backwards towards the east! Two promontories (West Sulawesi and South Sulawesi – the 'legs') project southwards, the 'tail' (East Sulawesi) points eastwards, and the long thin neck (North Sulawesi) at the north end of the island curves round into an arc facing south across the Molucca Sea. Sulawesi is interesting in being composed of several pieces of crust with quite different origins: westernmost Sulawesi is part of Sundaland, which by the Cretaceous formed part of Eurasia; South and East Sulawesi, the southern and eastern promontories, were originally part of continental Australia; while North Sulawesi is part of an active volcanic arc along the margin of the Celebes Sea Basin.

The eastern part of Indonesia is dominated by the large island of New Guinea, of which only the western half (Irian Jawa) belongs to the State of Indonesia;

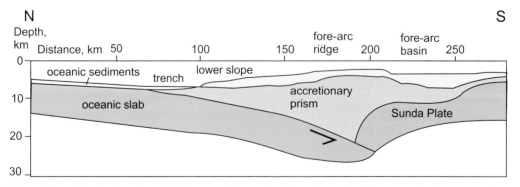

Figure 9.5 The offshore Sunda Arc. Simplified topographic section based on refraction seismic data across the western part of the arc, showing the accretionary prism. Lower higher-density part of the prism in orange; upper, lower-density material in yellow; oceanic sediments in blue. After Kopp, 2011.

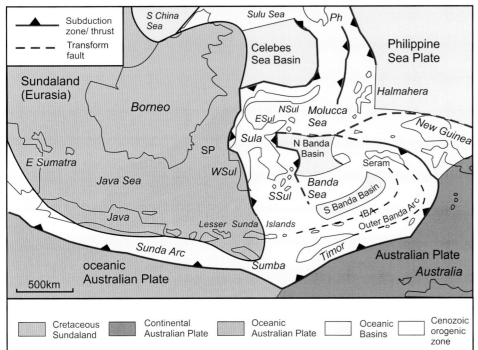

Figure 9.6 Main plate-tectonic elements of Central and Eastern Indonesia. 'Sundaland' represents the extent of stable Eurasian crust added in the Cretaceous to the northern side of the Cenozoic orogenic belt system. ESul, East Sulawesi; IBA, Inner Banda Arc; NSul, North Sulawesi; PH, Philippines; SSul, South Sulawesi; Sula, Sulawesi; WSul, West Sulawesi. After Hall, 2011.

the eastern half is the separate country of Papua New Guinea. New Guinea is situated on the northern side of the Australian continental plate, and belongs to the circum-Pacific orogenic belt system; it contains a central spine of high mountains, rising to over 5000m. This belt continues northwards through the island of Halmahera and continues into the Philippines. This sector of the belt will be discussed in the following chapter. There are several small ocean basins in the topographically complex region east of Sundaland. From north to south, these are: the Sulu, Celebes, North Banda and South Banda Basins. The Celebes Basin is bounded by a subduction zone on its eastern and southern sides; the North Banda Basin is bounded by a subduction zone on its western side and a transform fault to the north; the South Banda Basin is situated on the inner side of the Inner Banda Arc. On the eastern side of the orogenic belt is the Pacific Ocean, which is represented here by the Philippine Sea Plate.

Tectonic context

The region of Indonesia east of Sumba island, including Sulawesi, Timor, the western part of New Guinea, and the many smaller islands, is one of the most structurally complex of the whole Cenozoic orogenic belt system. The main reason for its complexity is its position at the junction of three plates: the continental part of the large Australian Plate to the south, Sundaland (part of Eurasia) to the west, and the oceanic Philippine Sea Plate to the east and north.

Figure 9.7 is a reconstruction of how the plate-tectonic framework of Eastern Indonesia evolved during the Neogene. During the Palaeogene Period, oceanic parts of two plates, the Indian and Australian Plates, were being subducted beneath Southeast Asia and the Philippine Sea Plate respectively, and the plate boundary network then was relatively simple. However, as the Australian continent approached Southeast Asia in the Miocene, as illustrated in Figure 9.7A, the shape of the northern margin of Australia became critical: the first point of contact was a large spur, the Sula Spur, which projected north-westwards from the western end of New Guinea. This NW–SE promontory contained the western part of New Guinea together with Seram, Southern and Eastern Sulawesi, and many smaller islands. Elsewhere, the oceanic part of the Australian Plate on the north-east side of New Guinea was being subducted beneath the Philippine Sea Plate to form the volcanic arcs of North Sulawesi, the Philippines and Halmahera, and the oceanic crust on the western side of the Australian Plate had not yet made contact with the Banda Arc.

By the late Miocene, Figure 9.7B shows that the Sula Promontory had bent into an east–west orientation, and broken up into several fragments during the convergent

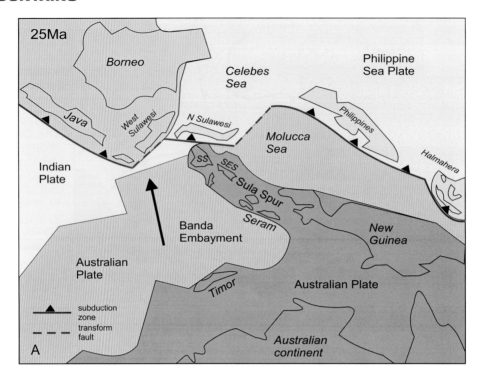

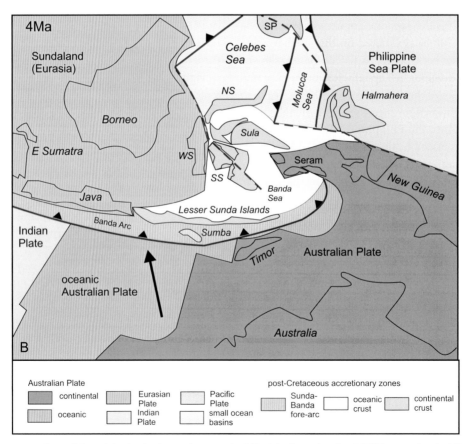

Figure 9.7 Reconstruction of the plate-tectonic environment of East Indonesia: **A**, at 25Ma ago (early Miocene) and **B**, 4Ma ago (early Pliocene). NS, North Sulawesi; SES, SE Sulawesi; SP, Southern Philippines; SS, South Sulawesi; WS, West Sulawesi. After Hall, 2011.

movements. The western oceanic part of the Australian Plate was now being subducted beneath the Sunda and Banda Arcs, and Timor, on the Australian continental shelf, was now in contact with the Banda Arc. The Australian oceanic crust on the northern, New Guinea, side had been subducted and this plate boundary became a transform fault. There now was a large area of complex deformation in a zone, coloured yellow in Figure 9.7B, between the active Banda Arc on its southern side, the Eurasian Sundaland Block on the west, and the ocean basins of the Pacific to the north. Here there is no well-defined boundary between the Eurasian and Australian Plates but a diffuse zone of fault blocks, much of which is submerged and obscured by recent sediments.

The Banda Arc

The two parallel arcs of the Banda Arc system form tight curves, convex towards the east, extending from Timor in the west to Seram and Buru in the north (Fig. 9.8), enclosing the Banda Sea. The inner, volcanic, arc extends from the island of Wetar at the western end of the arc, continues as a series of very small islands linked by a submerged ridge, and ends just south of the larger island of Seram. Nine volcanoes are situated

along the arc, most of which are currently active; three of the volcanoes are situated on the larger islands, but the others occur as separate volcanic mountains rising up to 5km directly from the floor of the Banda Sea. The outer, inactive, arc is situated along the Australian continental margin and extends from the large island of Timor in the southwest to Seram in the north.

The two arcs are separated by a fore-arc basin or trough, which varies in width from less than 50km in the south between the islands of Timor and Wetar, to over 150km in the centre of the arc where the ocean depth is over 7km. This reflects the fact that, at the island of Timor, the Australian continent is now in contact with Eurasia, the subduction zone is locked and active vulcanicity has ceased, whereas further north there is still some Australian oceanic crust that has not yet been subducted, and it is here that the vulcanicity is most intense.

The structure of the Banda Arc system is revealed both in Seram and Timor, where six tectonic zones can be distinguished, from northwest to southeast:

1 **_The Inner Banda Arc_**. This is the active volcanic arc.
2 **_The Fore-arc Basin_**. This extensional zone is a late Neogene basin, and contains the most recent Neogene deposits overlying the extended earlier strata.

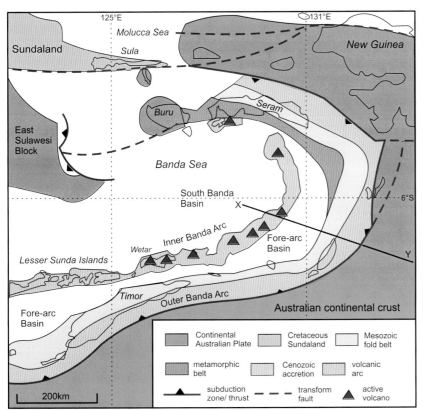

Figure 9.8 The Banda Arc system: main tectonic features. The inner Banda Arc is the presently active volcanic arc; the outer Banda Arc consists of an inner Mesozoic fold/thrust belt and an outer zone of Cenozoic accretion. The subduction zone defines the present surface boundary of the Australian continental plate, part of which has been subducted. After Darman, 2014

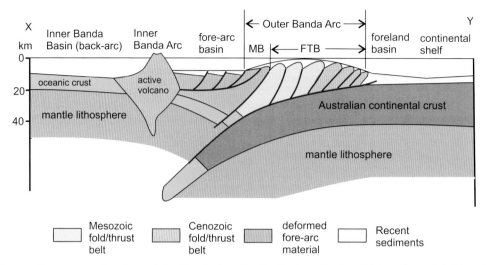

Figure 9.9 Schematic cross-section through the Banda Arc. Section is along the line X–Y on Figure 9.8. Note that the fore-arc basin is extensional and the outer arc compressional. FTB, Fold-thrust belt; MB, Ophiolite Belt. After Darman, 2014.

3 *The Ophiolite Belt*. Here, oceanic crust of the former Tethyan Ocean has been thrust northwards onto the passive margin of the Australian continent.

4 *The Metamorphic Belt*. This zone consists of low-grade schists derived from the Carboniferous to Jurassic sedimentary cover of the Australian continental platform. The zone wedges out southwards, and is absent at outcrop along the line of section of Figure 9.9.

5 *The Fold/thrust Belt*. This zone is divided into an inner part consisting of imbricated Cretaceous to Cenozoic deep-water sediments, and an outer of Mesozoic platform sediments, both derived from the Australian Plate.

6 *The Foreland Basin*. This zone contains a thick pile of recent sediments on the outer edge of the Australian platform.

Tectonic interpretation

The Outer Banda Arc, as seen for example on Timor and Seram, is an accretionary complex, where the older rocks are situated on the inner side of the arc, and the younger on the outer side (Fig. 9.9). The complex is interpreted as a set of successively older and deeper crustal slices thrust backwards up onto the Australian continent as it is subducted beneath the Eurasian Plate. The main phase of contractional deformation occurred in the late Miocene, when the main continental collision occurred. The structure has been imaged seismically across the line of section X–Y in Figure 9.8. The fore-arc basin here is extensional, which may be the result of

the late roll-back of the subducting slab. The volcanoes in the inner arc in this sector are shown rising directly from the ocean floor, which here is between 4km and 6km deep.

The origin of the Banda Sea Basin has been the subject of some debate: it has been interpreted as the result of back-arc spreading, in common with many of the arc systems around the Pacific rim; this would have occurred before the Australian continent came into contact with the arc. However, the shape of the arc must also have been influenced in part by the southward bending of the northwestern promontory of Australia as it came into contact with Eurasia, as illustrated by comparing Figures 9.7A and B. This would have resulted in the shape of the northern sector of the arc becoming much tighter, and assuming a more east–west orientation.

The Celebes and Sulu Seas

The most complex part of the Southeast Asian region of the Neogene orogenic belt is bounded in the west by Borneo; in the northeast by the island of Mindanao at the southern end of the Philippines Archipelago; and in the southeast by the island of Halmahera (Fig. 9.10). About 1150km from north to south and 850km across, it includes the strangely shaped island of Sulawesi discussed earlier, and the oceanic area to its north, which is divided into the Celebes Basin and the Sulu Basin, separated by the NE–SW-trending Sulu Ridge. These basins are bounded to the east by the Philippine Ridge, which continues south from Mindanao as a submerged feature to Halmahera. On the northwest side of the Sulu

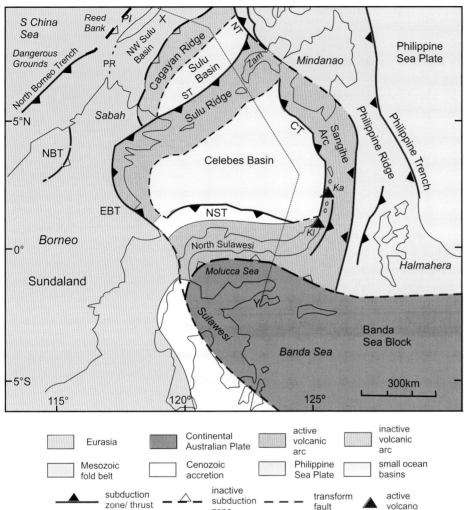

Figure 9.10 Main tectonic features of the Celebes Basin and surrounding area. CT, Cotabo Trench; EBT, East Borneo Thrust; Ka, Karangatan Volcano; KI, Klabat Volcano; NBT, N Borneo Thrust; NST, North Sulawesi Trench; NT, Negros Trench; PI, Palawan Island; PR, Palawan Ridge; ST, Sulu Trench; Zam, Zamboanga. After Rangin *et al.*, 1991.

Sea are two further ridges, the Cagayan Ridge and the Palawan Ridge, connecting the northeast end of Borneo (Sabah) to the Philippines.

The whole island of Sulawesi is mountainous: much of the main north–south central ridge of West Sulawesi is over 3000m in height. The most prominent peak on the ridge, Pantemano, is 3455m in height. This mountain ridge curves round into an east–west orientation along the narrow North Sulawesi peninsula, which also contains a central spine with several peaks over 2000m high. At its eastern end, this ridge continues offshore, bending round into a north–south-trending chain of small islands, the Sangihe Islands, situated along a submarine ridge that ends at southern Mindanao in the Philippines. There are thirteen volcanoes along this North Sulawesi–Sangihe Arc: nine on North Sulawesi and four on the Sangihe Arc. Klabat Volcano, at the northeast tip of North Sulawesi, is an impressive peak

rising to 2022m in height. However, the most active volcano in the whole arc in recent years has been Karangetang, on the small island of Api Siau, about 140km north of Mindanao.

The Sulu ridge is part of the Sulu volcanic arc, on the northwest side of the Celebes Basin. This mostly submarine ridge follows a chain of islands, the Sulu Islands, ending in the Zamboanga promontory in the Philippines to the northeast. At its southwest end, the Sulu Arc curves southwards through northeast Borneo. There is only one active volcano, Jolo, in the Sulu Islands chain, and a single volcano in Northeast Borneo, Bombolai, which may still be active, although there are no historical records of its eruption.

Another volcanic arc, situated on the northeastern side of the Sulu Basin, is known as the Cagayan Ridge. This ridge extends from northeastern Borneo to the island of Panay in the Philippines and is wholly submerged except for two small islands, both named 'Cagayan', at each end of the ridge. There is no active vulcanicity on this ridge now.

A fourth volcanic arc, now also inactive, lies further northwest and consists of the long island of Palawan together with a group of small islands to its northeast. This ridge lies within the South China Sea Block, which is part of the Eurasian Plate.

Plate-tectonic setting

This complex region lies at the intersection of three large plates: Eurasia to the northwest, the Philippine Sea to the east, and Australia to the south (Fig. 9.10). At its core are two oceanic basins: the Sulu Basin in the north and the larger Celebes basin to the south. At the eastern margin of the region, transecting both oceanic basins, is the Philippine Orogenic Belt, which is bounded on both sides by active subduction zones: the Philippine Trench on the eastern side, where the Philippine Sea Plate is being subducted beneath the Philippine Ridge; and the Negros and Cotabo Trenches on the western side.

The northern margin is the southeastern edge of the South China Sea Block, which forms part of the Eurasian Plate. To the west is the large island of Borneo, which by the Neogene also formed part of Eurasia, and to the south is the Banda Sea Block which, as described earlier, formed part of the Australian Plate until the beginning of the Miocene. Traversing this region from north to south, eight separate tectonic zones have been recognised: the South China Sea Block; the Palawan Ridge; the Northwest Sulu Basin; the Cagayan Ridge; the Sulu Basin; the Sulu-Zamboanga Ridge; the Celebes Basin; and the North Sulawesi–Sangihe Arc. Figure 9.11 gives a cross-section through these zones.

The South China Sea Block

This area, wholly submerged beneath the South China Sea, is known as the Dangerous Grounds in the south and the Reed Bank in the north. Drilling has revealed that the Mesozoic cover of the Eurasian Plate in the Reed Bank area had been disrupted by extensional block faulting during the Palaeogene, before being covered by undeformed Neogene sediments.

The Palawan Ridge

The Reed Bank Block is bounded on its southeastern side by the Palawan Trench, interpreted as the now inactive subduction zone along which the Eurasian passive margin formerly descended beneath the oceanic plate to its south. The Palawan Ridge, consisting of imbricated Palaeogene to early Neogene deposits, represents the accretionary prism formed during this subduction process.

The Northwest Sulu Basin

This narrow marine trough between the Palawan and Cagayan ridges marks the remnants of the oceanic basin that formerly separated the Reed Bank Block from the volcanic arc of the Cagayan Ridge.

The Cagayan Ridge

This mostly submerged ridge consists of andesitic and basaltic volcanic material of early Miocene age (c.22–15Ma) and represents a now extinct volcanic arc.

The Sulu Basin

This marine basin, interpreted as a back-arc basin to the Cagayan subduction zone, consists of oceanic crust of early Miocene age. On the southeast side of the basin is the Sulu Trench, along which oceanic crust of the Sulu Basin is being subducted beneath the Sulu Ridge. The Sulu Trench connects to the southwest with the

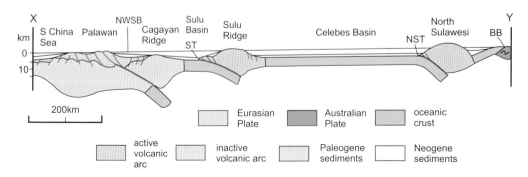

Figure 9.11 Simplified cross-section of the Sulu–Celebes area. Section is along the line X–Y on Figure 9.10. Note that the deformed sedimentary unit shown in blue also includes some early Neogene material. BB, Banda Sea Block; NST, North Sulawesi Trench; NWSB, Northwest Sulu Basin; ST, Sulu Trench. After Rangin *et al.*, 1991.

active East Borneo Thrust, and to the northeast with the Negros Trench, which forms part of the western boundary of the Philippine Ridge.

The Sulu–Zamboanga Ridge

This is a younger volcanic arc than the Cagayan, consisting mostly of volcanic material of Pliocene to Pleistocene age. However, there are only two active volcanoes on the ridge now, one (Jolo) on the Sulu Islands and a second on the Zamboanga peninsula in the Philippines. The Zamboanga sector of the arc also contains Miocene volcanics and is situated on pre-Cenozoic basement.

The Celebes Basin

The oceanic crust of the Celebes Basin is of mid-Eocene age or older, and the southeast side of the Sulu Ridge is interpreted as a passive margin. The magnetic anomaly pattern is oriented NE–SW, indicating that the crust of the marine basin spread southeastwards to be subducted beneath North Suluwesi. The basin may have originated as a result of back-arc spreading as the subduction zone at the margin of the Banda Sea Block retreated southwards, before the continental part of the Australian Plate made contact in the mid-Miocene.

The North Sulawesi–Sangihe Arc

This ridge consists of an accretionary prism resulting from the north-westwards subduction of the oceanic crust of the Molucca Sea prior to the mid-Miocene collision between Australia and Sundaland. The subduction zone fringing the ridge on its southern side is now inactive, but the North Sulawesi Trench on the northern side of the arc is currently active: volcanic activity took place in North Sulawesi during the early Miocene and a more recent phase commenced in the late Miocene. Three of the 13 volcanoes on the arc, including Karangetang, are considered to be still volcanically active.

Tectonic history

The Banda Sea Block collided with the southern margin of the Celebes Basin in Central Sulawesi at the beginning of the mid-Miocene (c.16Ma ago). Further convergence between Australia and Eurasia first induced rapid subduction along the Cagayan Trench in the north, resulting in the closure of the oceanic part of the South China Sea Plate and the development of the Cagayan Volcanic Arc. Between 15Ma and 10Ma ago, the Sulu Basin commenced subduction beneath the Sulu Ridge and was followed in the mid- to late Miocene by the southwards subduction of the Celebes Basin beneath the North Sulawesi–Sangihe Arc. Both oceanic basins are still being subducted along the Cagayan and Sulu Trenches. These basins are also being subducted eastwards beneath the Philippine Ridge via the Negros and Catabo Trenches, while on their western side, the subduction zones terminate against a thrust in Sabah, and a transform fault through central Sulawesi.

10

The Western Pacific Rim

The circum-Pacific orogenic belt, often called the 'Pacific Ring of Fire', consists of two distinct halves: the western, much of which is represented by oceanic islands and island arcs, and the eastern, which follows the western margins of the North and South American continents for most of its length, and is discussed in the following chapters. The southern end of the western circum-Pacific belt is at the southwest extremity of the South Island of New Zealand, and the orogenic belt proceeds from there along a complex series of arcs and island chains to New Guinea, the Philippines and Taiwan (Figs. 10.1, 10.2). From Taiwan, the Ryukyu Island Arc connects to Japan, which in turn links up with the Kamchatka Peninsula via the Kurile Island Arc. The last part of

Figure 10.1 Volcanic arcs of the Western Pacific Rim. © Shutterstock, from NASA image.

the chain is the Aleutian Arc, which connects Siberia with Alaska. A separate branch of the system forms a wide loop through the Mariana Island Arc, separating the Philippine Sea from the Pacific. Older parts of the orogenic belt are also present on mainland Asia, but are now separated from the present-day active belt by a series of back-arc basins.

The geological history of the circum-Pacific belt commenced with the supercontinent of Pangaea in the Permo-Triassic, when southern and eastern Asia were bordered by the ancestral Pacific Ocean Plate. Throughout the later part of the Mesozoic Era, various continental blocks broke away from Gondwana and were added to the Asian margin. Several of these, such as Cimmeria, North China and South China, have already been discussed in the preceding chapters. A succession of back-arc basins then formed during the Cenozoic, mainly as a result of trench roll-back, giving rise to the present-day system of arcuate Neogene belts, isolated from the continental margin.

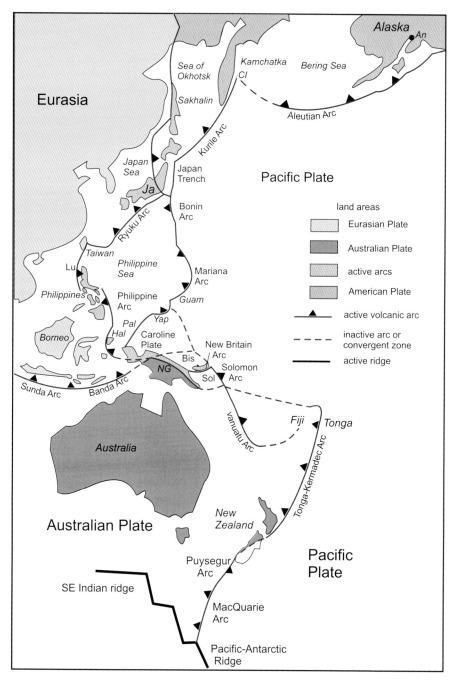

Figure 10.2 Active arc systems of the Western Pacific Rim. An, Anchorage; Bis, Bismarck Plate; CI, Commander Islands; Hal, Halmahera; Ja, Japan; Lu, Luzon Arc; NG, New Guinea; Pal, Palau; Sol, Solomon Plate. After Leat & Larter, 2003.

The New Zealand Sector

The Cenozoic mountain belt of New Zealand's South Island forms an impressive chain of mountains, 500km long and up to 160km wide, known as the Southern Alps, which form the central spine of South Island. The highest peak in the range, Mount Cook, is 3753m in height. The Southern Alps belt forms the western boundary of the Pacific Plate, and the southern tip of South Island links to the East Pacific ocean ridge via the MacQuarie Ridge, formerly regarded as a transform fault, although there is now thought to be a limited degree of convergence across it (Fig. 10.2). The mountain belt continues northwards through North Island as a greatly reduced narrow chain, also 500km long but much narrower, continuing the NE–SW trend to the end of the East Cape Peninsula.

Plate-tectonic context

New Zealand sits astride two tectonic plates: the Australian Plate to the west and the Pacific Plate to the east (Fig. 10.3). The plate boundary on South Island is the Alpine Fault, which runs along the west coast from Fjordland in the southwest, northeastwards to the north end of the island, where it divides into several branches. The main branch exits South Island north of the City of Christchurch, and continues northwards along the east side of North Island as a subduction zone, the Hikurangi Trench.

The southern end of the Alpine Fault is replaced by another subduction zone, the Puysegur Trench, which continues southwards along the west side of the mostly submerged Macquarie Ridge for about 400km. Along this sector of the plate boundary, it is oceanic crust of the Australian Plate that is being subducted, whereas further south along the Macquarie Arc, the Pacific Oceanic crust is being subducted beneath the Australian Plate. The Macquarie Arc meets the East Pacific mid-ocean ridge at the southwestern corner of the Pacific Plate.

Along the Hikurangi Trench, oceanic crust of the Pacific Plate is being subducted beneath North Island, which belongs to the continental Australian Plate. At the northeast end of North Island, the subduction zone continues as the Kermadec Trench, where oceanic Pacific Plate is being subducted beneath the Kermadec Arc. Both the Kermadec and North Island sectors are volcanically active: the six islands of the Kermadec group are volcanic, and the arc also contains many submarine volcanoes. North Island contains the Taupo volcanic zone, which is responsible for the famous hot springs and geysirs of Rotorua and the Tongariro volcanoes.

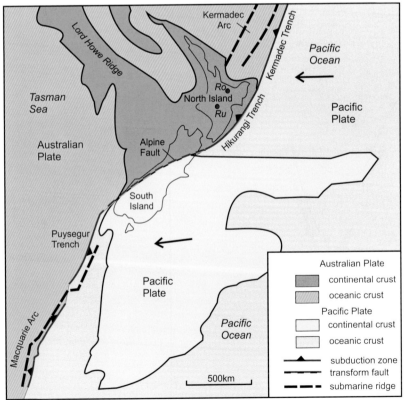

Figure 10.3 Plate-tectonic setting of the New Zealand region. Note that the plate boundary between the Australian and Pacific plates is a subduction zone east of North Island and also to the north and south of New Zealand, but is a transform fault through South Island with a component of convergence across it. After Coates, 2002.

Tectonic sub-division

The islands of New Zealand may be subdivided into five distinct tectonic units: the West Coast Province; the Volcanic Arc; the Older Torlesse Zone; the Younger Torlesse Zone; and the late Cretaceous–Cenozoic cover (Fig. 10.4).

The West Coast Province occupies a narrow belt, less than 100km in width, along the west coast of South Island, on the western side of the Southern Alps. It is bounded by the Alpine Fault along most of its length until the fault leaves the coast in Fjordland. South of this point, the rocks of the West Coast Province make up the southernmost part of the mountain chain. This tectonic zone consists of Precambrian and Lower Palaeozoic basement rocks of Gondwanan origin belonging to the Australian Plate. These Australian basement rocks also underlie North Island and continue westwards as a submarine platform beneath the Tasman Sea to link up with northeastern Australia. The New Zealand portion of this basement region rifted away from the Australian margin during the Cretaceous as the Tasman Sea basin opened up.

The Volcanic Arc

Immediately east of the west coast basement is a belt of volcanic rocks, including ophiolites, formed during Carboniferous to Jurassic times. In South Island, this zone is present only in the south and northwest of the island; between it is cut out by dextral displacement on the Alpine Fault. In North Island it continues as a 100km-wide belt along the west coast as far as Aukland. The volcanic arc was situated offshore along the eastern side of Gondwana, and resulted from the subduction of oceanic crust belonging to the ancestral Pacific Plate.

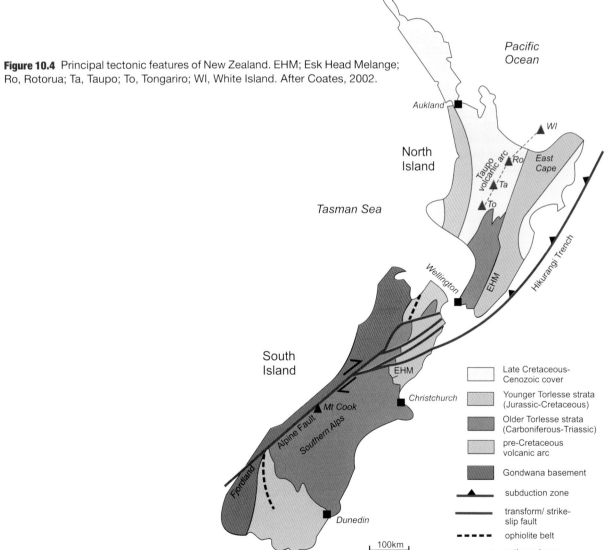

Figure 10.4 Principal tectonic features of New Zealand. EHM; Esk Head Melange; Ro, Rotorua; Ta, Taupo; To, Tongariro; WI, White Island. After Coates, 2002.

The Older Torlesse Zone

The Older Torlesse suite of rocks occupies the whole of the central part of South Island east of the Alpine fault, from Dunedin in the south to Christchurch, from where they extend in a narrow belt nearly to the north coast. They also occupy a smaller area in the southern part of North Island. The suite consists of a sequence, several tens of kilometres thick, mainly of greywackes and associated deep-marine sediments of Permo-Triassic age; these were deposited in submarine fans by turbidity currents flowing down the continental slope off the eastern margin of Gondwana. The western part of these Older Torlesse strata, close to the Alpine Fault, was uplifted to expose the metamorphosed lower parts of the sequence, known as the Haast Schist, during the First Rangitata Orogeny in the mid-Jurassic.

The Younger Torlesse Zone

The younger Torlesse suite, which occupies the northeastern part of South Island and the eastern part of North Island, consists of very similar sedimentary rocks to the older suite, but ranges in age from mid-Jurassic to mid-Cretaceous. They are separated from the Older Torlesse Zone by a narrow belt of disrupted sedimentary rocks known as the Esk Head Mélange. This belt represents a tectonic discontinuity formed when the older and younger parts of the accretionary complex were thrust together during the Second Rangitata Orogeny in the mid-Cretaceous.

The post-orogenic cover

Un-deformed sedimentary and volcanic deposits ranging from late Cretaceous to Recent form small sedimentary basins in South Island and cover a large part of North Island. Lavas and associated pyroclastic deposits originating from the active volcanoes of the Taupo Volcanic Arc are an important feature of North Island.

Tectonic History

Subduction along the eastern margin of Gondwana from late Carboniferous to Triassic times gave rise to a thick sequence of Older Torlesse strata, which were deposited on oceanic crust, then compressed and uplifted into a landmass during the First Rangitata Orogeny in the mid-Jurassic. A further period of accumulation of deep-marine strata ended with the Second Rangitata Orogeny in the mid-Cretaceous, which resulted in the formation of a wide accretionary complex encompassing almost all of New Zealand, together with much of the adjoining continental shelf. In the later Cretaceous, widespread extension and rifting resulted in the opening of the Tasman Sea basin, and the isolation of the various island landmasses that eventually became New Zealand.

During the Neogene Kaikoura Orogeny, subduction along the Hikurangi and Kermadec Trenches resulted respectively in the Taupo and Kermadec volcanic arcs. As can be seen in Figure 10.3, the convergence direction across the Hikurangi Trench is E–W, whereas on South Island, it is ENE–WSW, only slightly oblique to the Alpine Fault, which therefore is a transpressive structure with a large component of dextral strike-slip motion, responsible for the 500km displacement of the margin of the volcanic arc outcrop. The convergence rate is about 30mm/a in the south, varying to over 40mm/a in the north. Both North and South Islands are subject to intense and often highly destructive seismicity. Shallow earthquakes occur almost everywhere, whereas deep-focus earthquakes are concentrated on North Island and northernmost South Island due to the west-dipping Hikurangi subduction zone.

New Zealand to New Guinea

Northwards from New Zealand, the circum-Pacific orogenic belt lies entirely within the Pacific Ocean, through a series of intra-oceanic volcanic arcs, until it reaches New Guinea (Fig. 10.2). For the first 1000km it follows the submerged Kermadec ridge to the Tonga Archipelago, after which it turns sharply south-westwards to join the southern end of the Vanuatu Arc, and from there via the Solomon and New Britain Arcs to reach the northeastern coast of New Guinea.

The Tonga–Kermadec Arc

The Kermadec Arc is the northern extension of the Taupo volcanic zone of New Zealand's North Island, discussed previously. The arc includes six volcanic islands in the Kermadec Group together with about 30 submarine volcanoes, or seamounts. Only two of the volcanoes are known to be currently active. The volcanic ridge is accompanied by a trench on its eastern side, which reaches depths of up to 8km. The 800km-long Tonga sector of the Arc consists of 16 islands situated on a broad ridge containing 16 volcanoes, 12 of which are known to have been active historically (Fig. 10.5). On the eastern side of the ridge is the Tonga Trench, the southern end of which reaches a depth of nearly 11km. To the west of the Tonga Ridge is the older, inactive,

Lau Ridge, formed during Eocene to Miocene time, which has a more north–south trend and ends at the Fiji Platform. The two ridges are separated by the narrow, elongate, Lau back-arc basin, which opened over the last 6Ma, after which the Lau Ridge ceased to be active.

Despite the fact that for much of their lengths these volcanic arcs only occasionally break the surface, they are no less impressive as mountain ranges than their continental counterparts. The ocean floor from which they rise is typically between 2000m and 4000m in depth, and, measured from the deep-ocean trenches, a vertical relief of over 10,000km to some of the higher volcanic summits is not uncommon, making these systems comparable in scale to the Himalayas.

Tonga to Vanuatu

The Tonga Trench curves around the northern end of the Tonga Ridge, at the northern end of the Lau Basin (Fig. 10.5). It is replaced by the sinuous Vitiaz Trench Lineament, now inactive but active from Eocene to Miocene times, which forms the boundary between the Pacific Plate to the north and the North Fiji back-arc basin to the south.

The Lau Ridge ends on the eastern side of the Fiji Platform, on which the Fijian Islands are situated. Fiji itself consists of two large and several smaller islands, and the Fiji Platform is a large area, now part of the Pacific Plate, approximately 450x200km in extent, with an ocean depth of less than 500m. The crust here is estimated to be c.20km in thickness, which has prompted speculation that the platform may represent a thinned piece of the Australian continent, now separated from it by the South Fiji Basin. Alternatively, Fiji may have acquired an unusually large volume of igneous material from the opposed subduction zones on either side.

The south side of the actively spreading North Fiji Basin is defined by the wide curve of the Hunter Fracture Zone, which links the north end of the Lau Ridge with the southern end of the Vanuatu Arc. The submerged Hunter Ridge north of, and parallel to, the Hunter Fracture Zone is believed to be the site of subduction of oceanic crust generated in the North Fiji Basin in the last 5Ma or so, but the fracture zone itself appears now to be a transform fault, being approximately parallel to the northeasterly convergence direction of the Australian Plate.

The Vanuatu Arc

The Vanuatu Ridge is a broad, submerged volcanic ridge 1500km long and up to 200km wide, and the Vanuatu Island chain is spread out over 1300km along it (Fig. 10.5). The main NW–SE ridge is Y-shaped and divides in the central part of the archipelago, a more NNW–SSE branch continuing northwards to include the Santa Cruz Islands, which are part of the Solomon group. The former name for Vanuatu is the New Hebrides, and the trench that follows the ridge on its western side is still referred to as the New Hebrides Trench.

The archipelago consists of 13 larger islands and over 60 smaller islands and rocky outcrops. Twenty-four volcanoes are recorded in the Vanuatu Arc, 12 of which have been active in recent years. The largest Island, Espiritu Santo, contains igneous rocks of Eocene to Miocene age resting on Gondwana basement. The ridge rises from an ocean depth of over 3km on its eastern

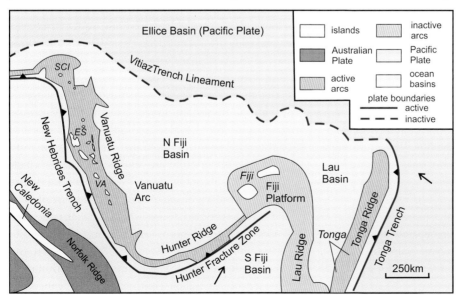

Figure 10.5 Plate-tectonic framework of the Tonga–Fiji–Vanuatu region. ES, Espiritu Santo; SCI, Santa Cruz Islands; VA, Vanuatu archipelago. After Segev *et al.*, 2012.

side and over 5km in the New Hebrides Trench to the west; the highest point on it is the 1810m-high volcanic peak of Tabwemasana on Espiritu Santo.

The arc is strongly active seismically due to the ongoing subduction of oceanic crust belonging to the Australian Plate beneath the North Fiji Basin. Convergence rates are in the range 90–120mm/a, except in the central sector, where they are much reduced due, it is thought, to the intersection of the trench with a submerged ridge. The effect of this concurrence appears to have caused the uplift of the outer ridge and the creation of a narrow basin between the outer and inner arcs east of Espiritu Santo.

The Solomon Arc

At the northern end of the Vanuatu Ridge is a small group of islands, the Santa Cruz Islands, which are part of the Solomon Islands (Fig. 10.5). From there, the seismically active belt bends abruptly into an east–west orientation. The trench here is known as the San Cristobal, or Solomon, Trench, and is accompanied on its northern side by a 200km-wide ridge containing the Solomon Islands archipelago (Fig. 10.6). This island group is an independent country consisting of six main islands and over 900 smaller ones extending over a distance of about 1500km. One of the largest islands is Guadalcanal, site of the famous Second World War battle.

There are three separate sectors of the Solomon Arc, each resulting from the subduction of different plates. The southeastern sector, containing the island of Guadalcanal, is the product of the subduction of the northward movement of oceanic crust of the Australian Plate beneath the Pacific plate. The central sector, centred on New Georgia, is the site of subduction of younger, more buoyant crust of the Woodlark Plate. Here the trench is indistinguishable as a topographic feature, and this part of the ridge exhibits lower relief. The northwestern sector is where oceanic crust of the Solomon Sea Plate is being subducted; the trench here is again well defined and the ridge contains the large, mountainous island of Bougainville. There are eight recorded volcanoes on the arc, but only four have been active historically.

New Guinea

This island consists of the two separate countries of Irian Jaya in the west and Papua New Guinea in the east, but is usually referred to as 'New Guinea' for geological purposes.

The plate-tectonic arrangement between the Solomon Arc and New Guinea is complex, with four separate small plates intervening between the Pacific Plate in the north and the Australian Plate in the south (Fig. 10.6). The Solomon Ridge continues westwards through the large island of New Ireland and the numerous small islands of the Bismarck Archipelago. Much of this broad, arcuate ridge on the north side of the Bismarck Sea is submerged, and it dies out westwards. It is the main component of the North Bismarck Plate and is defined on its northern side by a currently inactive plate boundary marking the former position of the circum-Pacific subduction zone. Its southern boundary consists partly of spreading axes and partly of transform faults, and is seismically active.

The South Bismarck Plate consists partly of oceanic crust belonging to the Bismarck Sea Basin and partly of the continental crust of New Britain and northeastern New Guinea. To the south of the South Bismarck Plate is the actively spreading Solomon Sea Plate. This oceanic plate is being subducted northwards beneath the South Bismarck Plate along the well-defined New Britain Trench, which is over 9km deep in places.

All three of these micro-plates are bounded along their southern sides by the long, narrow Woodlark Plate, which contains the mountainous spine of New Guinea. This belt of high mountains, containing several peaks over 4000m in height, divides eastwards into three separate ridges: a northern arm, which continues as the New Britain Arc; a central arm, which forms the Woodlark Ridge and contains the small Woodlark Island; and a southern arm, which continues the mountain spine along the southeastern promontory of Papua New Guinea and continues as the submarine Louisiade Ridge. The northern and central ridges are both southerly-convex arcs enclosing the Solomon Sea Basin. The southern arm is part of the Australian Plate.

The Woodlark Plate consists mostly of material accreted to the Australian margin during the Palaeogene. The central mountain range consists of Mesozoic metamorphic rocks folded and uplifted as a result of a collision between the Australian continental margin and a Miocene volcanic arc, and is fringed by a belt of obducted ophiolites along its northern side. The northern boundary of the Woodlark Plate east of New Guinea is the Woodlark Ridge, which forms the southern boundary of the Solomon Sea Plate, and is seismically active.

The southern half of the island of New Guinea is composed of Australian continental crust consisting

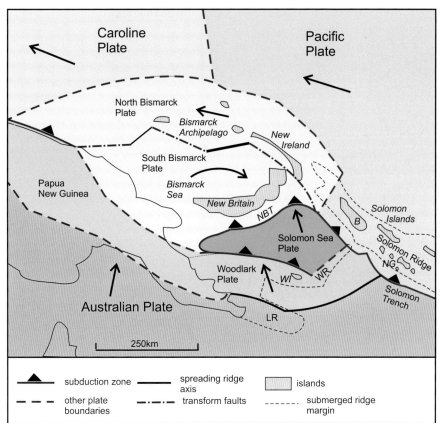

Figure 10.6 Small plates between the Solomon Ridge and Papua New Guinea. B, Bougainville; LR, Louisiade Ridge; NBT, New Britain Trench; NG, New Georgia; WI, Woodlark Island; WR, Woodlark Ridge. After Alataristarion 2006, via Wikimedia Commons.

of Precambrian basement together with the remnants of a Palaeozoic orogenic belt. The boundary between the Australian and Woodlark Plates on New Guinea is inactive, but its continuation eastwards is considered to be a spreading axis separating the two plates.

The Philippines

The circum-Pacific orogenic belt turns northwards from western New Guinea through the island of Halmahera and along the Philippine Ridge to the southern end of the Philippines (see Fig. 10.2). The Philippines consist of over 7000 islands extending in a roughly north–south belt over 2000km long and up to 500km wide along the Western Pacific rim (Fig. 10.7). The two largest islands are Luzon in the north, and Mindanao in the south. Both these islands contain mountain ranges with many peaks over 2000m in height, the highest being the 2954m volcanic peak of Mt. Apo in Mindanao. The smaller islands of Mindoro and Panay in the centre of the archipelago are also mountainous. However, most of the mountain groups are quite isolated, and there is no clearly defined mountain spine running through the island chain.

The Philippines is one of the most volcanically active sectors of the circum-Pacific belt, with about fifty active volcanoes and seamounts. These are arranged in two main arcs: a northern, running parallel to the Manila Trench subduction zone on the western side of the archipelago, and a southern, parallel to the Philippine Trench on the eastern side. Mayon (Fig. 10.8) in Luzon is the most famous of the active volcanoes. There are also groups of active volcanoes on the island of Negros, associated with subduction along the Negros Trench, and in southwestern Mindanao, belonging to the Cotabo Arc.

Tectonic summary

The Philippine Islands are framed by subduction zones on each side: the Manila, Negros and Cotabo Trenches on the western side and the East Luzon Trough and Philippine Trench on the eastern (Fig. 10.7). Only the Philippine Trench is a major topographic feature, being over 10.5km deep at its deepest point. Both sets of subduction zones dip inwards beneath the Philippines, which is also split down the centre by a major sinistral strike-slip fault system. The combination of E–W compression and sinistral strike-slip indicates that the

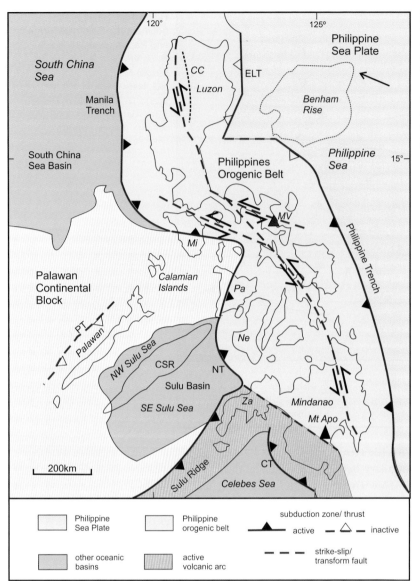

Figure 10.7 Tectonic framework of the Philippines. BR, Benham Rise; CSR, Cagayan-Sulu Ridge; CT, Cotabo Trench; ELT, East Luzon Trough; MT, Manila Trench; NT, Negros Trench; PT, Palawan Trough. Geographic names: CC, Cordillera Central; Mi, Mindoro; MV, Mayon volcano; Ne, Negros; Pa, Panay; Za, Zamboanga. After Morrison, 2014.

Figure 10.8 Mayon Volcano, Luzon. This 2462m peak is the most famous of the active volcanoes of the Philippines and last erupted in 2014. © Shutterstock, by suronin.

Philippine belt is essentially transpressional: the Philippine Plate is converging with Eurasia in a northwesterly direction with an estimated rate of convergence along the Philippine subduction zone of 150mm/a.

The Philippines commenced its geological history as a volcanic arc along the Asian margin during the Cretaceous. Subsequent retreat of the subduction zone led to the development of the extensional South China Sea Basin. The southern part of the South China Sea Basin consists of the mostly submerged Palawan continental block, containing the NE–SW ridge of Palawan and the Calamanian Islands. This block has impinged on the western margin of the Philippine belt, causing a prominent bend in the subduction zone in the central part of the archipelago.

The southern part of the Philippine Belt is complicated by the intersection of the active Sulu volcanic arc with the western margin of the belt. Here oceanic crust of the Sulu Basin is being subducted eastwards beneath the Negros Trench as well as southeastwards beneath the Sulu Ridge. Another ocean basin centred on the Celebes Sea, south of the Sulu Ridge, is being subducted beneath the Cotabo Trench west of Mindanao. The tectonic history of the Celebes and Sulu Basins, together with the inactive Palawan and Cagayan arcs, was discussed in the previous chapter (see Figs 9.10 and 9.11).

The sinuous shape of the Philippine Belt and the associated deviation of the strike-slip fault system are attributed to the impingement of the buoyant Palawan block in the central part of the belt. This has caused anti-clockwise rotation of the islands on the north side of the projecting wedge and clockwise rotation of those on the south side. The boundary between the Palawan Block and the Philippine Belt is represented by a fold-thrust belt, exposed on Mindoro and other islands to the immediate south of it, in which the Philippine Belt rocks are overthrust towards the west over the Palawan Block.

The eastern margin of the Philippine Belt is simpler; the northern sector is bounded by the East Luzon Trough, which is now an inactive subduction zone. This zone is offset along a transform fault to continue as the active Philippine Trench.

The basement of the Philippine Belt is similar to that of the Palawan Block and consists of Cretaceous volcanics and ophiolites originating on the Asian margin. These are overlain by Eocene to Oligocene volcanics, carbonates and clastic sediments. The older rocks, which include metamorphosed basement units, are exposed in the Cordillera Central mountain range in western Luzon. The uplifted older rock outcrops are separated by basins containing Neogene sediments and volcanics.

Tectonic history

Volcanism began along the Asian margin during the Cretaceous, forming a volcanic arc which then split from the Asian margin with the opening of the South China Sea Basin in the Eocene. At about the same time, the Philippine Sea Plate separated from the Pacific Plate. Later in the Eocene, the spreading South China Sea Plate began to subduct eastwards beneath the Philippines along the Manila Trench, forming a Palaeogene volcanic arc. The Celebes Sea Basin also opened during this period.

During the Miocene, the Palawan Block collided with the western margin of the belt, causing the northeastern end of the block to underthrust the Philippine Belt on Mindoro and Panay. Subduction was temporarily halted along the western margin of the belt, but later in the Miocene, the Sulu Sea Basin opened, subduction began along the Cotabo and Negros Trenches, and resumed along the other sectors, continuing into the Pliocene. However, at present, active subduction appears to be restricted to the Manila, Philippine, Negros and Cotabo sectors.

The Luzon Arc and Taiwan
The Luzon Arc

The circum-Pacific orogenic belt continues northwards from Luzon in the Philippines for 400km along the Luzon Volcanic Arc to meet the southern end of Taiwan (see Fig. 10.2). There are five active volcanoes along the southern sector of the Luzon Arc, which occupies a poorly defined, mostly submerged, ridge becoming indistinct south of Taiwan. The east-dipping Manila Trench subduction zone, where Eurasian crust is descending beneath the Luzon Ridge, defines the western side of the ridge but is not readily apparent topographically.

The East Luzon Trough is a more prominent topographic feature that follows the eastern side of the ridge up to the southern promontory of Taiwan. This west-dipping subduction zone marks the plate boundary between the Philippine Sea and Eurasian Plates, and is replaced in eastern Taiwan by an east-dipping onshore thrust (Fig. 10.9).

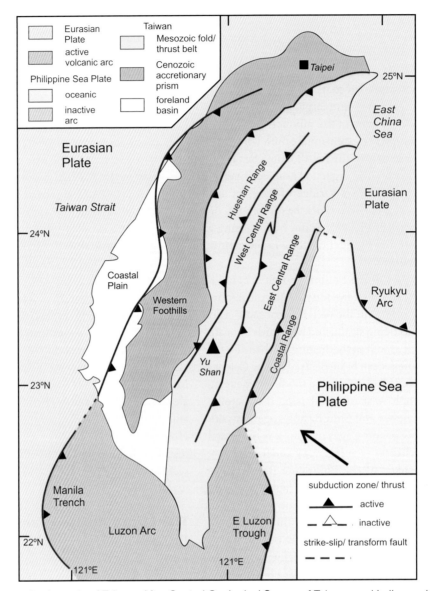

Figure 10.9 Main tectonic elements of Taiwan. After Central Geological Survey of Taiwan; and Lallemand *et al.*, 2002.

Taiwan

The greater part of the island of Taiwan is composed of a central mountainous spine, the Taiwan Shan, which runs the whole 380km length of the island. Most of the mountain range is over 2000m in height, the highest peak being Yu Shan (Fig. 10.10), at 3997m.

The island can be divided into four main tectonic zones, from west to east: the Western Coastal Plain; the Western Foothills; the Central Highlands; and the (eastern) Coastal Range. The three western zones are part of Eurasia, whereas the Coastal Range is part of the Philippine Sea Plate (Fig. 10.9). The Western Coastal Plain is the stable margin of the Eurasian Plate and is the site of the present-day foreland basin. The Western Foothills represent the uplifted synorogenic accretionary prism, the western margin of which is an east-dipping thrust marking the edge of the Coastal Plain.

The Central Highlands consist of three separate thrust packages: the Hueshan, West Central and East Central Ranges, containing a series of thrust, folded and metamorphosed units involving Pre-Cenozoic basement together with an Eocene to Miocene sedimentary cover.

The Coastal Range is the on-land continuation of the Luzon Arc, and is separated from the Central Highlands by a major east-dipping thrust representing the boundary

Figure 10.10 Yu Shan (Mt. Jade): the highest mountain in Taiwan. Shutterstock©elwynn.

between the Eurasian and Philippine Sea Plates. The convergence rate along this boundary is estimated at 80mm/a in a WNW–ESE direction. About two-thirds of the way along the eastern coast of Taiwan, this plate boundary turns abruptly eastwards, and then southeastwards to become the Ryukyu Arc, discussed below. Subduction of oceanic crust has ceased beneath the main island of Taiwan and the stretched Eurasia continental crust is in contact with the now inactive Luzon Arc along the eastern Coastal Range. Convergence there is being accommodated along the set of thrusts that separate the main tectonic units. Except for some submarine volcanoes offshore, volcanic activity is now limited.

The Japanese Arc System

There are five separate elements making up the Japanese arc system. The main circum-Pacific orogenic belt includes, from south to north: the Ryukyu Arc, the SW Honshu Arc, the NE Honshu Arc and the Kuril Arc (Fig. 10.11). The fifth element is the Izu-Ogasawara, or Bonin, Arc, which meets the main belt in Central Honshu. The present configuration was established during the early Neogene when the opening of the Japan Sea separated the Japanese islands from Eurasia.

The Ryukyu Arc

This is the southernmost sector of the system, extending from Taiwan to Kyushu, a distance of 1000km. It consists of about 55 islands, some quite large – up to several tens of kilometres in length – and hilly, while others are flat coral atolls. The largest island is Okinawa, in the centre of the chain. The larger islands are all volcanic; there are five currently active volcanoes, all in the central and northern sectors of the chain.

The Ryukyu Arc in its present form dates back to the late Pliocene, about 2Ma ago, as a result of the subduction of the Philippine Sea Plate westwards beneath Eurasia. The arc consists of a partly submerged ridge, convex towards the Philippine Sea, the southern part of which is composed of deformed Mesozoic and Cenozoic basement similar to that of Taiwan, while the northern part is volcanic in origin. The ridge is bounded on its eastern side by the Ryukyu Trench, which is a marked topographic feature with a maximum depth of 7460m. The rate of subduction along the trench has been calculated at $c.52$mm/a. There is a much broader, but shallower, linear basin on the inner side of the ridge, known as the Okinawa Trough, which is interpreted as an extensional back-arc basin produced by slab roll-back of the subduction zone.

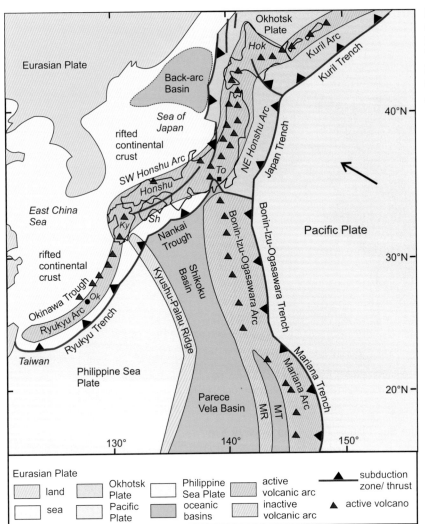

Figure 10.11 Main tectonic elements of the Japan arc system. MR, Mariana Ridge; MT, Mariana Trough. Geographic names: Hok, Hokkaido; Ky, Kyushu; Ok, Okinawa; Sh, Shikoku; To, Tokyo. After Taira *et al.*, 2016.

The SW Honshu Arc

This arc extends for about 600km from the southern tip of the island of Kyushyu, through SW Honshu and Shikoku Island, to just west of Tokyo, where the Izu-Ogasawara ridge meets the coast. This part of Japan is hilly, with some peaks just short of 2000m in height along the southern side of the arc in Kyushu and Shikoku. The on-land sector of the arc here is around 200km in width. There is a single active volcano on the inner side of the arc; the outer part of the arc, including Shikoku Island, is interpreted as a fore-arc ridge.

The basement of the SW Honshu Arc consists of a series of accretionary prisms and thrust slices dating from Palaeozoic to Cenozoic in age, becoming younger south-eastwards. The arc is bounded on the outer side by the Nankai Trough, where oceanic crust of the Shikoku Basin, part of the Philippine Sea Plate, is being subducted beneath the arc. The Shikoku fore-arc consists of a series

of basins separated by thrusts. On the inner side of the arc, the Japan Sea back-arc basin is situated on rifted Eurasian continental crust. Palaeomagnetic evidence has shown that the opening of this basin was achieved by the clockwise rotation of SW Honshu coupled with anti-clockwise rotation of NE Honshu.

The central part of Honshu, north of the city of Tokyo, is much more mountainous, with several ranges containing peaks over 3000m in height. This elevated area, where the SW and NE Honshu Arcs meet, is thought to result from a collision with the Izu-Ogasawara Ridge, and marks the western boundary of the Okhotsk Plate.

The NE Honshu Arc

This is over 700km long and extends in a NNE–SSW direction from Central Honshu to the northern tip of the island of Hokkaido. It contains two roughly parallel lines of Pleistocene to Present-day volcanoes, 17 of which are currently active. This sector of the arc consists of a basement similar to that of the SW Honshu Arc, with thrust packages separated by N–S-trending reverse faults. The on-land arc is bounded on the eastern side by the Japan Trench, along which the Pacific Plate is being subducted at a rate of 80–90mm/a. This region is extremely active seismically and there have been a number of very large earthquakes in recent years, including the notorious 2011 episode responsible for the damaging tsunami that destroyed the nuclear power plant at Fukushima.

Tectonic structure of the SW Honshu Arc

Prior to the opening of the Japan Sea back-arc basin in the Miocene, the basement of the main Japanese islands formed part of the Asian mainland, and consists of a series of accretionary complexes attached to the Asian margin from late Palaeozoic times. The basement of the SW Honshu Arc is divided into a number of tectonic zones, each of which represents a composite accretionary package (Fig. 10.12). These are mostly separated by thrusts or steep faults, the most important being the Median Tectonic Line (MTL), which is a long-lived sinistral strike-slip fault active from Cretaceous times to the present day.

The tectonic zones are of three basic types: unmetamorphosed or weakly metamorphosed accretionary complexes; high-pressure, low-temperature metamorphic belts and low-pressure, high-temperature metamorphic belts. These adjoining high-pressure and low-pressure metamorphic belts are the 'paired metamorphic belts' made famous by Akiho Miyashiro in the 1960s (Miyashiro, 1961), and were subsequently explained as the product of, respectively, deeply buried slabs of cool subducted crust, and sections of warm volcanic-arc crust. Paired belts developed during the Jurassic and again in the Cretaceous.

There are seven roughly parallel tectonic zones in the SW Honshu Arc (Fig. 10.12). On the inner (northwest) side of the MTL, from north to south, are: the Hida basement; the Suo High-Pressure Zone; the Mino-Tamba complexes; and the Ryoke Low-Pressure Zone. South of the MTL are: the Sanbagawa High-Pressure Zone; the Chichibu Complex and the Shimanto Complex.

The Hida Zone

This metamorphic terrain consists of granitic gneisses and amphibolite-facies metasediments ranging in age from Precambrian to Jurassic and considered to represent a block of Eurasian basement affected by mid-Mesozoic deformation and high-temperature metamorphism. The Hida rocks have been correlated with the Sino-Korean Block on the Asian mainland.

The Suo Zone

This high-pressure, low-temperature metamorphic zone consists of schists interpreted as a Permo-Triassic accretionary complex metamorphosed during the Jurassic. The Suo belt is restricted to the southern part of Honshu and Kyushu and is thrust beneath the overlying Mino-Tamba units.

The Mino-Tamba accretionary complexes

The wide zone identified as 'Mino-Tamba' in Figure 10.12 contains several thrust units in addition to the Mino-Tamba (*sensu stricto*), including the Akiyoshi and Ultra-Tamba accretionary complexes and the Maizuru ophiolite complex, and consists of oceanic sediments and volcanics ranging from Carboniferous to Jurassic in age and interpreted as Permian to Jurassic accretionary complexes. The zone is also represented in NE Honshu and Hokkaido, where it is known as the North Kitakami Zone, and can be traced northwards into the Sikhote-Alin Range on the Russian mainland.

The Ryoke Zone

This is a mid-Cretaceous high temperature–low pressure metamorphic belt developed from the accretionary complex of the Mino-Tamba Zone and has a gradational boundary with it. The equivalent units in NE Honshu and Hokkaido are known as the Abukuma and Hidaka Belts respectively.

The Sanbagawa Zone

This is a high-pressure–low temperature metamorphic unit that has experienced high strain due to its proximity to the Median Tectonic Line. It consists of oceanic crust and trench deposits of early Mesozoic age. The Sanbagawa and Ryoke are 'paired' metamorphic belts developed during the mid-Cretaceous event.

The Chichibu Zone

This is another accretionary complex similar to the Mino-Tamba units, containing sediments ranging from Silurian to Permian in age. It is separated by a fault from the Shimanto Zone on the south side but has a gradational boundary with the Sanbagawa Zone.

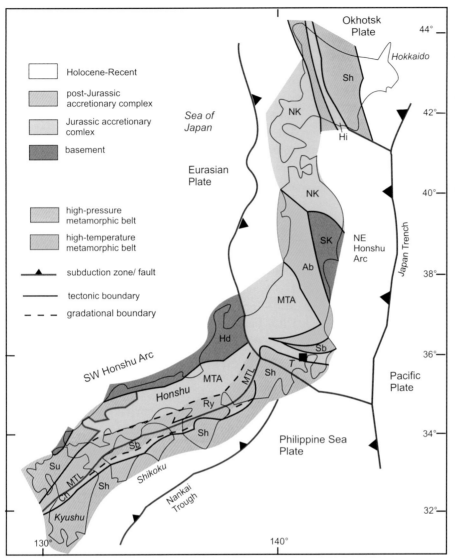

Figure 10.12 Tectonic zones of the Honshu Arcs. Low-P/High-T metamorphic zones: Ab, Abukuma; Ry, Ryoke. High-P/Low-T zones: Sb, Sanbagawa; Su, Suo. Accretionary complexes: Ch, Chichibu; Hi, Hidaka; MTA, Mino-Tamba; NK, North Kitakami; Sh, Shimanto. Foreland blocks: Hd, Hida; SK, S. Kitakami; MTL, Median Tectonic Line. T, Tokyo. After Kojima *et al.*, 2016.

The Shimanto Zone

This is an accretionary complex of Cretaceous to Neogene age, which developed on the southeast side of the earlier accretionary complexes. A similar zone in Hokkaido is considered to be an extension of the same belt.

Tectonic history of the Honshu Arcs

The basement of the Honshu Arcs contains rocks ranging from mid-Precambrian to mid-Palaeozoic in age, similar to the Sino-Korean Block of the Asian mainland. From the Permian onwards, a succession of accretionary complexes developed as a result of the subduction of the Pacific oceanic plate beneath the Asian margin. The earlier packages (e.g. Mino-Tamba and Chichibu) were deformed and metamorphosed during the early

Cretaceous Oshima orogeny, and were succeeded by a later complex (the Shimanto) which developed through the Cretaceous and Palaeogene and was deformed in a second event (the Mizuho orogeny) in the Miocene. The current phase of activity extended from the Miocene to the present and is responsible for the active volcanic arc and the Nankai accretionary prism developing on the outer side of the arc.

The Okhotsk Plate and the Verkhoyansk Belt

The Okhotsk Plate (or sub-plate) lies at the southwestern corner of the American Plate but has a slightly different movement direction and convergence rate from the main part of that plate. Its western boundary is the subduction zone lying west of NE Honshu, and extending northwards along the west side of Sakhalin Island. From there, the plate boundary links with the Verkhoyansk Fold Belt, which is a compressional fold belt formed by the late-Jurassic to early Cretaceous collision between the North American and Eurasian Plates (see Figs 10.1 and 10.2).

The Bonin-Mariana Arcs

This arc system (see Figs 10.1 and 10.11) extends from central Honshu southwards for 2800km to the island of Guam, where it is joined by two shorter arcs around the islands of Yap and Palau, eventually meeting the long north–south submarine ridge that ends at the western end of New Guinea. The northern (Japanese) sector is also known as the Izu-Ogasawara Arc. The arc system consists of a ridge with a deep trench on its outward side, and represents an intra-oceanic convergent margin where the Pacific Plate is being subducted westwards beneath the Philippine Sea Plate. The northern Bonin, or Izu-Ogasawara, Trench is over 10km deep near Honshu but becomes less obvious further south. The Marianas Trench, at over 11km deep, is one of the deepest parts of the global trench network.

The Bonin–Mariana arc system originated about 50Ma ago and resulted in a simplification of the circum-Pacific boundary, isolating the Philippine Sea on its inner side. From about 25Ma ago, back-arc extension opened up the Parece Vela Basin as the arc migrated eastwards, separating it from its original site along the submarine Kyushu–Palau Ridge. The northern part of this back-arc basin is now being subducted beneath SW Honshu along the Nankai Trough. The northern sector of the arc system is a broad submerged ridge, up to 200km wide, with only a few small islands, but hosts seven active volcanoes. The Mariana sector is more arcuate in shape, and convex towards the Pacific. A more recent back-arc basin, the Mariana Trough, has formed on the inner side of this arc, splitting the ridge into two parts. A further five active volcanoes occur in this southern sector.

The Kuril–Kamchatka Arc

This arc extends from the northeastern end of Hokkaido Island for about 1000km to the Kamchatka Peninsula in Russia (see Fig. 10.2). The island arc consists of a string of islands on a submerged ridge with a marked trench, the Kuril Trench, up to 10.5km deep on its outward side. The Sea of Okhotsk on the inner side of the arc is a Neogene back-arc basin. Vulcanicity on the Kuril arc commenced about 20Ma ago in the early Miocene, and there are about 40 volcanoes active at present. The Kamchatka sector of the arc contains another 29 active volcanoes whose activity commenced in the late Pliocene, and contains Kluchevskoi Volcano, at just under 5000m Eurasia's highest active volcano.

The arc meets the Aleutian Arc at a triple junction west of the Commander Islands, where a transform fault continues northwards and the Aleutian Arc strikes eastwards. The Pacific Plate is being subducted north-westwards beneath the Kuril Arc at a rate of between 75mm/a in the south and 83mm/a in the north. The oblique nature of the convergence means that the stress along the arc is transpressional, with a sinistral strike-slip component.

The Aleutian Arc

This is a broad arc, about 3400km long, convex towards the Pacific Ocean, with the Bering Sea Basin on its inner side. The arc extends from its junction with the Kuril–Kamchatka Arc to Alaska; the last 1000km lie along the Alaskan Peninsula to the city of Anchorage, where it meets the northern end of the Western American orogenic belt system – the subject of the next chapter.

The arc is situated on a broad ridge containing 14 large islands together with a large number of smaller ones. There are at least 76 volcanoes on the arc, of which 36 are known to have been active historically, and large earthquakes are frequent. The Pacific Plate is being subducted beneath the arc in a northwest direction at a rate of about 80mm/a. The Aleutian Trench on the outer side of the arc is up to 7800m deep in its central sector, but becomes shallower and less marked at the western end of the arc, where the convergence direction is highly oblique; the arc here is effectively a transform fault.

The North American Cordillera

The circum-Pacific orogenic belt continues eastwards from the Alaska Peninsula to join the mountain ranges of Northern Alaska. These lie at the northern end of the North American Cordillera, a series of mountain ranges that occupy the western side of North America, stretching from Alaska in the north to Mexico in the south, a total distance of around 7800km (Figs. 11.1, 11.2).

The Alaska–Canada sector is up to 1000km wide and consists of several separate mountain ranges including the Brooks and Alaska Ranges in northern Alaska, the Mackenzie Mountains in the Yukon, the Coast Mountains in British Columbia, and the Rocky Mountains on the eastern side of the Cordillera, which extend from British Columbia into Alberta (see Figure 11.2 for province and state boundaries). The highest summits are in the coastal sector; Denali (formerly Mt. McKinley) in the Alaska Range, at 6190m, is the highest peak in North America (Fig. 11.3). The eastern ranges are less high; the peaks of the Canadian Rockies are generally around 3000m.

Further south, through the USA, the Cordillera broadens to over 1500km, and is more clearly divided into the Cascade Range and Sierra Nevada on the western side, and the Rocky Mountains of Wyoming and Colorado on the eastern side. Between these ranges is the vast volcanic region of the Columbia Plateau and, further south, the Basin and Range Province, characterised by a series of mountain ranges with intervening valleys created by extensional faulted blocks. This whole region of the western USA is elevated above 1500m. The highest peak in the coterminous United States is Mount Whitney, at 4421m, in the Sierra Nevada.

The Cordillera continues southwards through Mexico, narrowing to about 200km in width towards the border with Guatemala. The main mountain range here is the Sierra Madre Occidental.

Figure 11.1 Mountain ranges of the North American Cordillera. McK Mts, McKenzie Mountains. Shutterstock©Arid Ocean.

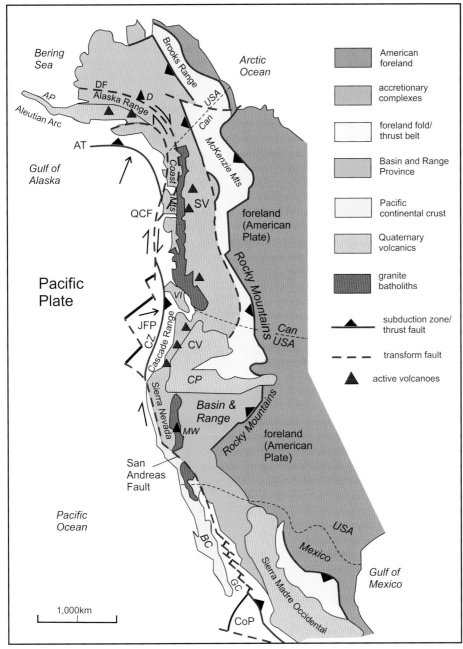

Figure 11.2 Principal tectonic features of the North American Cordillera. AP, Alaska Peninsula; AT, Aleutian Trench; BC, Baja California; CoP, Cocos Plate; CP, Columbia Plateau; CV, Cascade volcanic belt; CZ, Cascade subduction zone; DF, Denali Fault; GC, Gulf of California; JFP, Juan de Fuca Plate; MW, Mt Whitney; QCF, Queen Charlotte Fault; SV, Stikine volcanic belt; VI, Vancouver Island. After Moores & Twiss, 1995.

Plate-tectonic overview

This sector of the circum-Pacific orogenic belt system (Fig. 11.2) is superficially rather simpler than the western Pacific sector. The American Plate forms the eastern foreland, and the Pacific Plate, together with several smaller oceanic plates, lies on its western side. Between, however, is a complex region consisting of a large number of separate terranes and tectonic provinces.

The plate boundary in the northern, Alaska, sector is the Aleutian Trench, which sweeps round from an east–west trend at the eastern end of the Aleutian Arc to north–south, parallel to the coast, at which point it becomes the Queen Charlotte transform fault. Near the northern end of Vancouver Island, the plate boundary bends into a NW–SE orientation and forms part of the boundary of the Juan de Fuca Plate (the surviving part of the original Farallon Plate). The eastwards subduction of this oceanic plate along the Cascadia Subduction Zone, which lies off the coasts of Washington and Oregon,

Figure 11.3 Denali (Mount McKinley) in the Alaska Range. At 6190m, this is the highest mountain in North America. ©
Shutterstock, by bcambell65.

is responsible for the many active volcanoes of the
Cascades Range.

At its southern end, the Cascadia subduction zone
meets the famous San Andreas transform fault, which
runs for 2000km through Western California and into
the Gulf of California, where it meets the ridge forming
the western boundary of the Cocos Plate. The northern
end of this ridge is chopped into small segments by a
series of short transform faults. West of the San Andreas
Fault is a long, thin sliver of continental crust belonging
to the Pacific Plate and encompassing the coastal region
of Western California and the Baja California Peninsula
of Mexico. The eastern boundary of the Cocos Plate
is formed by another subduction zone, along which
this oceanic plate is being subducted beneath Central
America. This part of the Circum-Pacific belt is the
subject of the next chapter.

The Alaska–Canada Sector

This part of the circum-Pacific orogenic belt was where
the terrane concept was first developed by Coney and
his co-workers in 1980. Since then a large number of
separate terranes have been proposed, some of which
are regarded as 'super-terranes', themselves consisting of
more than one individual terrane. Many of the terranes

are regarded as 'displaced' or even 'exotic' because they
display evidence of having travelled some distance from
their original site of formation. The system of nomen-
clature and interpretation of the terrane complex is to
some extent in a state of flux because the region is still
comparatively poorly known and difficult of access.

The main part of the Alaska–Canada sector is composed of a number of displaced terranes, which have
been added to the North American Plate during the
Mesozoic. The main collision event took place during
the Cretaceous Period. The terranes represent a mixture
of crystalline metamorphic complexes, volcanic arc
material, arc–trench deposits of various ages, and ophiolite units containing oceanic sediments and ultramafic
rocks. Some of these can be shown to have originated
far from their present position and are believed to have
been transported northwards on the Farallon plate and
scraped off the subducting slab.

For the purpose of this account, the region is described in terms of seven terranes or super-terranes,
plus several autochthonous tectonic zones, based on
the subdivision recognised by the Geological Surveys of
the Yukon and British Columbia (Fig. 11.4). The whole
complex forms a wide arc oriented roughly east–west in
Alaska, and bending round to trend NW–SE in British

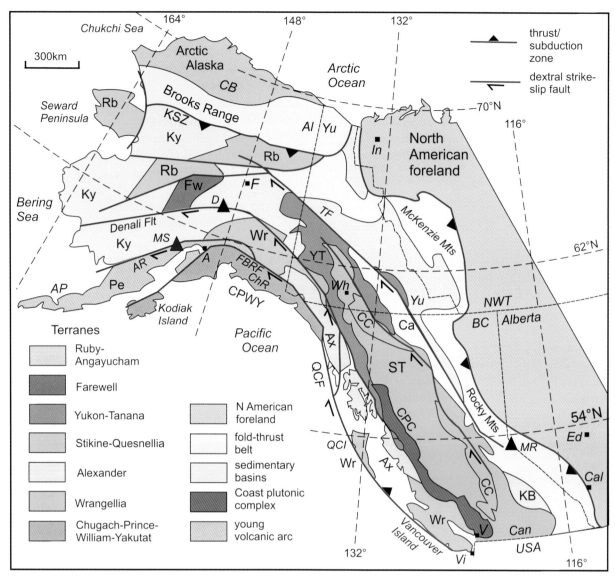

Figure 11.4 The Alaska–Canada sector of the North American Cordillera. AP, Alaska Peninsula; Ca, Cassiar belt; CB, Colville Basin; CPC, Coast Plutonic Complex; FBRF, Fairweather-Border Ranges Fault; KB, Kootenay Basin; KSZ, Kobak suture zone; Ky, Koyukuk, Nyak, Togiak Basins; QCF, Queen Charlotte Fault; TF, Tintina Fault. Terranes: Ax, Alexander; CC, Cache Creek; CPWY, Chugach–Prince William–Yakutat; Fw, Farewell; Pe, Peninsular; Rb, Ruby–Angayucham; Wr, Wrangellia. Cities: A, Anchorage; Cal, Calgary; Ed, Edmonton; F, Fairbanks; In, Inuvik; V, Vancouver; Vi, Victoria; Wh, Whitehorse. States/ Provinces: Al, Alaska; BC, British Columbia; Can, Canada; NWT, Northwest Territories; Yu, Yukon. Mountains: AR, Alaska Range; ChR, Chugach Range; D, Dinali; MR, Mt Robson; MS, Mt. Spurr (volcano). After Geological Surveys of the Yukon and British Columbia, 2011.

Columbia. On the north or east side of the belt is the North American foreland and foreland fold-thrust belt.

The foreland

The North American foreland consists of the ancient Precambrian core of North America, known as Laurentia, or 'the Canadian Shield' in Canada, plus material added to it during the Palaeozoic. The Alaskan sector of the foreland lies within the Arctic Alaska Terrane.

The Arctic Alaska Terrane

This 400km-wide zone in Northern Alaska contains rocks ranging in age from Precambrian to Cenozoic, representing the basement and sedimentary cover of the North American foreland. The southern part of the terrane, occupied by the mountains of the Brooks Range, is a foreland fold-thrust complex, while the northern part is known as the Colville Basin. The northern boundary of the zone on land is the north Alaskan coast, but the zone continues offshore onto the continental shelf. This whole region north of the Brooks Range is well known to petroleum geologists as the 'North Slope'. The southern boundary of the terrane is defined by the Kobuk suture zone, which marks the original margin of the American Plate. This suture zone is a south-dipping thrust along which rocks of the Angayucham Terrane (see below) have been overthrust northwards over the Arctic Alaska Terrane, forming the highest nappe of the fold-thrust belt.

The foreland fold-thrust belt

This belt extends eastwards from the coast of the Chukchi Sea, following the mountains of the Brooks Range in Arctic Alaska across the Alaska–Yukon border towards the boundary with the Northwest Territories, where it is cut off by a strike-slip fault. South of this fault, the belt continues through the Mackenzie Mountains and into British Columbia; here it is represented by the Canadian Rocky Mountains, which extend south-eastwards through Alberta and into the USA. The peaks of the Canadian Rockies are generally around 3000m, the highest being Mt. Robson, at 3954m (Fig. 11.5).

The sedimentary cover in this belt consists of Palaeozoic to Mesozoic platform and continental-slope deposits resting on the Precambrian basement of the Canadian Shield. These strata record the history of the passive margin of the North American continent that existed until subduction of the Farallon plate commenced at its western margin in the Mesozoic. The belt is dominated by folded and thrust units of this sedimentary cover, and provides one of the most spectacular and best documented examples of thrust tectonics.

Foreland and intermontane basins

Several large sedimentary basins overlie the margin of the North American Plate. The Colville Foreland Basin, north of the Brooks Range in Arctic Alaska, contains Upper Cretaceous to Cenozoic sedimentary rocks derived from the rising mountains of the Brooks Range. Another, much wider, intermontane basin lies west of the Mackenzie Mountains in Northwest Territories, but narrows southwards through British Columbia where it dies out. This basin is bounded on its west side by a major dextral strike-slip fault, the Tintina Fault. In southern Alberta, this basin reappears as the Kootenay Basin.

Figure 11.5 Mount Robson, Jasper National Park. This 3952m peak is the highest mountain in the Canadian Rocky Mountains. © Shutterstock, by BGSmith.

The central accretionary complex

This zone is about 800km wide in Alaska but narrows to around 600km through the British Columbian sector. It consists of three main fault-bounded blocks that are themselves split into sections by fault splays. Each block forms a northerly convex arc trending from NE–SW in the west to NW–SE in the southeast. Each of these blocks is overlain by Cretaceous–Cenozoic sedimentary basins at their western ends in Alaska. The faults bounding these blocks have all experienced a dextral strike-slip sense of movement, which has resulted in a general anticlockwise rotation of the more southerly blocks relative to the American foreland, and caused an accumulation of terranes at the western end of the complex. The fault blocks themselves do not correspond to terranes, since each block contains more than one terrane, and the boundary faults cut across the terrane boundaries in places, since the main fault-block movements post-date the terrane accumulation.

Seven separate tectonic units are differentiated in Figure 11.4, between the foreland thrust belt and the coast. Some of these are composite 'super-terranes' and all of them are regarded as displaced to a greater or lesser extent from their origins. The terranes are, from northeast to southwest: Ruby–Angayucham, Farewell, Yukon–Tanana, Stikine–Quesnellia, Wrangellia, Alexander and the composite Chugach–Prince William–Yakutat unit.

Ruby–Angayucham

This is a composite super-terrane consisting of a Precambrian–Palaeozoic basement (Ruby) of North American foreland type overlain by allochthonous oceanic rocks of Permo-Triassic age (Angayucham) deformed and metamorphosed in the Upper Cretaceous Laramide Orogeny. The unit forms two main outcrops, on the Seward Peninsula and north of the city of Fairbanks. It also forms a narrow band along the southern edge of the Arctic–Alaska Terrane, where it has been overthrust onto the latter terrane within the Brooks Range. The central outcrop is overlain on its northern, western and southern sides by sedimentary basins that received terrigenous sediments ranging from the Upper Cretaceous onwards.

Farewell

This terrane is regarded as an exotic continental block. Its contacts with neighbouring terranes are obscured by the sedimentary basins that surround it on three sides, and it is separated by a fault from the Ruby–Angayucham unit to the north.

Yukon–Tanana

This unit consists of Upper Palaeozoic and older rocks, thought to belong originally to the continental margin of the North American foreland, which adjoins it to the east. The unit has been transported north-westwards by dextral movements along the Tintina Fault. Its southwest boundary is the dextral Denali Fault.

Stikine–Quesnellia

This composite terrane represents a Permian to Jurassic volcanic arc, or arcs, accreted to the North American continent during Jurassic subduction, and includes two large outcrops of ophiolitic rocks, known as the Cache Creek assemblage. The terrane adjoins continental rocks of the Cassiar fold-thrust belt to the east, while the Yukon–Tanana terrane lies on its southwest side in the northern half of the outcrop, and the Alexander and Wrangellia terranes in the southern. The intrusions of the Coast Range Plutonic Complex occupy a 1100km-long strip within the southeastern part of the outcrop.

Wrangellia

This composite super-terrane is an assemblage of island arcs and oceanic plateaux of Carboniferous to Jurassic age considered to have formed in an equatorial region far to the south of its present location, and transported northwards on the Farallon Plate. The terrane forms a large outcrop in Alaska, sandwiched between the Denali and Border Ranges faults and another in southern British Columbia, where it occupies the Queen Charlotte Islands and Vancouver Island. Here, the Queen Charlotte Fault forms the western boundary and the Alexander Terrane the eastern.

Alexander

This terrane consists of marine shales, carbonates and red beds of Silurian to Devonian age containing faunas similar to those of North-eastern Siberia. The terrane is considered to have amalgamated with Wrangellia during the Carboniferous and it has been suggested that Wrangellia and Alexander should be considered as a single terrane. The Alexander rocks have been correlated with those of the Farewell Terrane, discussed above.

The Aleutian volcanic arc

This currently active volcanic arc extends through the Alaskan Peninsula and into the Alaska Range, north of Anchorage in southeastern Alaska, and is the result of the northward subduction of the Pacific Plate. There are more than 80 potentially active volcanoes in Alaska, 43 of which have been active historically. Most are on the

Aleutian island arc, discussed in the previous chapter. One of the more recently active volcanoes in mainland Alaska is Mount Spurr, 130km west of Anchorage, whose 1992 explosive eruption deposited ash as far as the city itself.

The Coast Range Arc

The Coast Range plutonic complex, often termed the Coast Range Batholith, extends for 1100km from Southern Alaska to Vancouver, parallel to the coast and up to 100km inland. It consists of a variety of late Cretaceous to Eocene plutonic bodies of typically granodioritic composition, intruded into the volcanic rocks of the Stikine Terrane. The arc, produced by the subduction of the oceanic Farallon Plate, is a typical example of an 'Andean' type of active continental margin. Once this plate had been consumed beneath the continental margin, the relative motion of the Pacific and American plates was parallel to the coast, and the plate boundary became a transform fault. The volcanic arc continues southwards into the Cascade Range of Washington State in the USA, where it is still active (see below).

The outboard terranes

These consist of three fault blocks, arranged in an arc along the southern coast of Alaska. The largest, and innermost, the Chugach Terrane, extends in an arc from Kodiak Island through the Chugach Mountains and wedges out near the Yukon border. It is bounded to the north by the Border Ranges Fault, on which the terrane is overthrust northwards over the Aleutian volcanic arc. The terrane reappears as a coastal sliver further south, where it is in fault contact with the Alexander Terrane. The two smaller fault blocks, the Prince William and Yakutat terranes, occur on the outer side of the Chugach terrane, along the southern Alaska coast. These terranes collectively represent an accretionary prism developed on the outer side of Wrangellia during the Mesozoic subduction phase. Since subduction ceased during the Cenozoic, the outboard terranes have been moving northwards relative to the North American continent along the Fairweather–Border Ranges Fault at a rate currently estimated at c.45mm/a.

Tectonic history

Throughout the Palaeozoic, the eastern part of the orogenic belt lay along the passive margin of the North American continent. Subduction of the Farallon Plate, part of the ancestral Pacific system, along the western margin of the continent commenced in the Jurassic. Several of the individual terranes, including Wrangellia and Alexander, accreted together first before colliding with the American continent in the mid-Cretaceous. This collision was responsible for the main phase of deformation and metamorphism within the orogen and is known as the Laramide Orogeny. The process of terrane accretion is illustrated schematically in Figure 11.6.

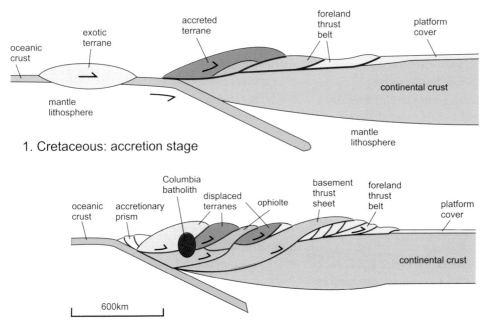

1. Cretaceous: accretion stage

2. Palaeogene: prior to strike slip displacement

Figure 11.6 Cartoon sections showing the Canadian sector before strike-slip displacement. How the orogen could have been constructed by successive accretions of displaced terranes. Note that the convergence direction was probably highly oblique to the continental margin. Colours as in Figure 11.4.

The later, Cenozoic, history of the orogenic belt was characterised by lateral movements along several major dextral strike-slip faults, parallel to the Queen Charlotte and San Andreas transform faults that form the present Pacific–American plate boundary. These movements slid the various terranes northwards relative to the eastern part of the orogen. Late-orogenic collapse due to gravitational spreading resulted in extensional faulting that has reversed the movement on some of the earlier thrust faults. The northwestern part of the orogen hosts a Cenozoic accretionary prism and parts of the coastal belt are still tectonically active.

The Western United States and Mexico

After crossing into the United States, the Cordilleran belt expands to over 1500km in width, taking in the four States of California, Nevada, Utah and most of Colorado at its widest part. This section of the Cordillera can be divided into three main tectonic zones and/or provinces: the Rocky Mountain belt (including the Colorado Plateau), the Basin and Range Province and the western accretionary zone, together with four igneous provinces: the Columbia Volcanic Province, the Cascade Volcanic Arc, the Sierra Nevada Batholith and the Sierra Madre Occidental Volcanic Province (Fig. 11.7).

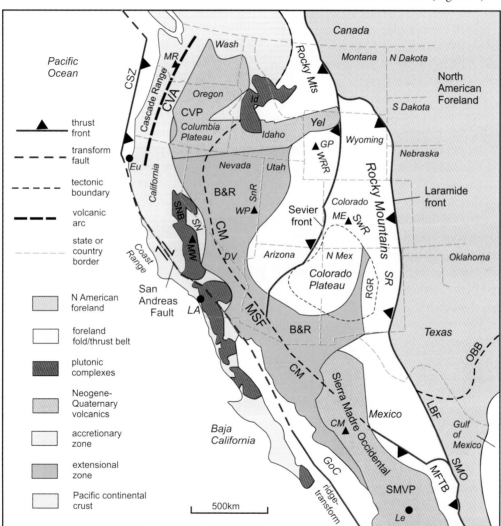

Figure 11.7 Principal tectonic features of the Western USA and Northern Mexico. B&R, Basin and Range; CM, continental margin (pre-orogenic) ; CSZ, Cascades Subduction Zone; CVA, Cascades Volcanic Arc; CVP, Columbia Volcanic Province; LBF, La Babia Fault; MFTB, Mexican fold-thrust belt; MSF, Mohave–Sonora Fault; OBB, Ouachita Belt boundary; RGR, Rio Grande Rift; SMVP, Sierra Madre Occidental Volcanic Province; SNB, Sierra Nevada Batholith. Geographic names: CM, Cerro Mohinora (mountain); DV, Death Valley; GoC, Gulf of California; GoM, Gulf of Mexico; GP, Gannett Peak; ME, Mt. Elbert; MR, Mt. Rainier; MW, Mt. Whitney; SMO, Sierra Madre Oriental; SN, Sierra Nevada; SnR, Snake Range; SR, Sacramento Range; SwR, Sawatch Range; Wash, Washington; WP, Wheeler Peak; WRR, Wind River Range; Yel, Yellowstone. After Coney et al., 1980; and Fitz-Diaz et al., 2011.

The Rocky Mountains

Along the eastern side of the Cordilleran belt, the Rocky Mountains belt continues from Alberta into Montana, where it divides into eastern and western branches, each containing individual ridges over 3000m high, with numerous peaks over 4000m. The eastern branch, which continues through Wyoming and into Colorado, is divided into several distinct ranges. The highest peak in this sector is Mount Elbert (4401m) in the Sawatch Range in Colorado. This belt continues southwards, as the Sacramento Range, through New Mexico, to the Mexican border, after which it continues through northern Mexico as the Sierra Madre Oriental Range. The western branch forms a wide, eastward-facing arc through western Wyoming and Utah into Arizona. Gannett Peak (4202m) in the Wind River Range in Wyoming is the highest in this sector.

As in the Alaska–Canada sector, the Rocky Mountains belt corresponds to the marginal fold-thrust zone developed on the North American foreland. The eastern branch is known as the Laramide Belt, and the western, the Sevier. The Laramide Belt is characterised by uplifted fault-bounded basement blocks of the North American Foreland, together with their platform cover, whereas the Sevier is a typical thin-skinned belt, in which the folds and thrust sheets involve only the sedimentary cover, and the basement is not involved (Fig. 11.8). The eastern branch continues into Mexico, where it occupies the mountains of the Sierra Madre Oriental and is cut off by the thin-skinned Mexican fold-thrust belt.

The Sevier Belt represented the western margin of the North American continent until the Cretaceous, and its cover contains strata ranging from Cambrian to mid-Mesozoic in age. The deformation here began in the Jurassic and continued into the early Cretaceous, whereas the Laramide deformation took over in the late Cretaceous and lasted until the Eocene.

The Colorado Plateau

Between the Laramide and Sevier mountain belts is a region of relatively featureless ground, mostly over 2000m in altitude, including large parts of Colorado, New Mexico, eastern Arizona, and southeastern Utah, known as the Colorado Plateau. This is a largely desert area drained by the Colorado River and its tributaries, which cut through the horizontal Mesozoic strata that cover much of the region and provide the spectacular coloured cliffs seen in famous US national parks such as Glen Canyon and Grand Canyon (Fig. 11.9). To avoid confusion, it should be noted that what is described here, and defined in Figures 11.7 and 11.8, is the geological, or tectonic, province; however, the Colorado Plateau as described geographically extends much further west, and includes the southern part of the Sevier branch of the Rocky Mountain belt. The eastern part of the Plateau is split by the Rio Grande Rift, which is an extensional fault-bounded valley occupied by the Rio Grande river.

The Precambrian basement of the Colorado Plateau is part of the North American Foreland, and the platform cover contains strata dating from Cambrian to Neogene in age. The basement and its Palaeozoic cover are visible in the walls of the Grand Canyon, but much of the scenery elsewhere on the Plateau is formed from brightly coloured Mesozoic sandstones. The stability of

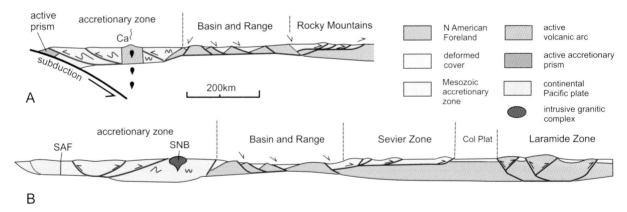

Figure 11.8 Schematic sections across the North American Cordillera: **A**, northern and **B**, southern sectors to illustrate the types of crustal structure. The northern margin is dominated by the actively subducting Cascade Arc, whereas the southern is bounded by the San Andreas Transform Fault. CA, Cascade Arc; Col Plat, Colorado Plateau; SAF, San Andreas Fault; SNB, Sierra Nevada Batholith. Note that individual structures are illustrative only and not intended to be accurate. After Moores & Twiss, 1995.

Figure 11.9 The Grand Canyon, Arizona, in the southwestern part of the Colorado Plateau. Note the horizontal strata. © Shutterstock, by Jason Patrick Ross.

the Plateau is the result of its relative strength, which has enabled the orogenic stresses arising from tectonic activity to the west to be transmitted though the region to the Laramide belt on its eastern side.

The Basin and Range Province

This wide region extends from Oregon and Idaho in the north to the Sierra Madre Occidental Volcanic Province in the south, and widens to about 800km between the Sierra Nevada Mountains in the west and the Sevier ranges in the east. The province is characterised by a series of narrow mountain ranges separated by wide flat valleys. The central and northern parts of the province are generally over 2000m in altitude, with the individual mountain ranges rising to over 3000m. Further south, in Arizona and northern Mexico, much of the land is less than 1000m high. The highest mountain is Wheeler Peak (3982m) in the Snake Range, in Nevada. The valleys are mostly hot dry deserts, including the notorious Death Valley near the western margin of the province.

The Basin and Range Province is well known to geologists as a classic example of extensional tectonics. The mountain ranges correspond to uplifted blocks

(horsts) bounded by normal faults, and the valleys to the intervening structural depressions (graben). The faults are listric: that is, they level out at depth to enable large extensional displacements to take place along them (Fig. 11.8). Many of the uplifted blocks expose the Precambrian basement and are known as metamorphic core complexes; such complexes are typical of extended regions and were discussed previously in the chapters on the Alpine–Himalayan Belt.

Although the crust is of normal thickness, the province experiences a higher than normal heat flow, and the mantle lithosphere is abnormally thin. This explains the relatively high altitude and structural weakness of the Basin and Range, which contrasts with the strong, cool lithosphere of the Colorado Plateau. The origin of the province as a distinct tectonic entity commenced with the low-angle subduction of the Farallon Plate in the Jurassic, which created an over-thickened crust in the whole Rocky Mountain belt up to the Laramide Front, including both the Basin and Range and the Colorado Plateau. After the mid-Eocene, the subduction regime changed: the convergence direction became oblique, and a volcanic arc formed along the western margin of the province, attributable to a steepening of the slab. This resulted in the initiation of a back-arc extensional regime on the upper plate, confined to the region overlying the subduction zone, but excluding the Colorado Plateau. Estimates of the amount of extension vary from about 80% upwards, but this amount could produce the required amount of thinning from an initial thickness similar to that of the adjacent Colorado Plateau. The extensional phase lasted until around 25Ma ago, when the subduction zone was replaced by the San Andreas Transform Fault.

The Western Accretionary Zone

This zone consists of displaced terranes accreted to the North American continent during the Mesozoic, similar to those described in the Alaska–Canada sector of the orogen (Fig. 11.7). The wide northern part of the zone hosts the Columbia and Cascade Arc volcanic provinces. Further south, in California, the zone narrows; much of it here is occupied by the Sierra Nevada Batholith.

At a point about 150km from the northern Californian border, near the city of Eureka, the plate boundary at the American continental margin changes from subduction zone to transform fault, with an accompanying abrupt change in the topography of the interior. From here southwards, the western side of the Cordillera consists of two parallel mountain ranges: the Coast Range, which hugs the Californian coast for over 900km nearly as far as Los Angeles, and the much more impressive Sierra Nevada, situated around 200km to the east and separated from it by a wide plain.

The San Andreas Fault divides the accretionary zone into two: the eastern side belongs to the North American Plate whereas west of the fault, the zone consists of displaced terranes belonging to the Pacific Plate.

The Columbia Volcanic Province

West of the Rocky Mountains, the northern part of the Cordillera in the USA is markedly different from its counterpart in Canada (Fig. 11.7). The Basin and Range Province extends northwards into Oregon, where much of the area is flat, relatively featureless desert – more basin than range – and eastwards into Idaho. Here, and further north, the Basin and Range Province is overlapped by the geographical region known as the Columbia Plateau, which in tectonic terms is part of the same extensional region as the Basin and Range.

The Columbia Plateau covers an area of over 500,000km^2, covering large parts of the States of Washington, Oregon and Idaho. Much of it is over 1000m elevation, and several peaks reach over 3000m in height. It is bounded on its western side by the mountains of the Cascade Range and in the east by the Rocky Mountains belt. Most of the Plateau is underlain by Neogene to Recent igneous rocks of the Columbia Volcanic Province, including the voluminous Columbia River plateau basalts, which are up to 3km thick and cover an area of around 160,000km^2.

The origin of the volcanism, which includes the active hydrothermal activity of Yellowstone National Park, is thought not to be related to the Cascade subduction zone, but has been attributed to the presence of a mantle hotspot currently situated beneath Yellowstone. The ages of the volcanics become younger eastwards, believed to be indicative of the westward movement of the North American Plate over a stationary hotspot. The region of the plateau in general must have experienced considerable extension and thinning in order to accommodate the lava pile.

The Cascade Volcanic Arc

The Cascade Range is a relatively narrow belt of volcanic mountains, situated about 150km from the coast, that extend from southern British Columbia for 1100km through Washington and Oregon into northern California (Fig. 11.7). The range hosts many peaks over 3000m in height, the highest being Mount Rainier, at 4392m, in Washington State. Only two volcanoes have been active in recent times – one being Mt. St. Helens, immediately south of Mt. Rainier, which is notorious for the eruption that produced the catastrophic mud avalanche of 1980. The northern, Canadian, sector is volcanically inactive. The Cascade Volcanic Arc is the result of the subduction of the Juan de Fuca Plate (part of the old Farallon Plate) along the Cascadia Subduction Zone, commencing in the late Eocene, and is all that remains of the active subduction system that followed the whole coast of North America during the Mesozoic (see Fig. 11.2).

The basement of the Cascade Arc consists of displaced terranes accreted to the North American continent during the Mesozoic, similar to those described in the Alaska–Canada sector of the orogen.

The Sierra Nevada Batholith

The Sierra Nevada Range is about 640km in length and 110km across, and includes several peaks over 4000m in height, including Mount Whitney (4421m), the highest peak in the United States south of Alaska (Fig. 11.10). Much of this range is carved out of the Sierra Nevada Batholith, which is an Upper Cretaceous plutonic complex formed at the time when the Farallon Plate was being subducted beneath the west coast, and similar to the Coast Plutonic Complex of Canada.

The Sierra Madre Occidental Volcanic Province

The mountains of the Sierra Madre Occidental form the central spine of Mexico from the US border to near the City of León, and form the western rim of the central Mexican plateau (Fig. 11.7). The range is everywhere over 2000m high, the highest peak being Cerro Mohinova, at 3300m. The volcanic province was initiated during the Eocene, after the Laramide event, due to the subduction of the Farallon Plate beneath the continental margin, at that time west of Baja California. This volcanic activity ceased in the late Eocene after the commencement of rifting in the Gulf of California, to be replaced in the south by the younger Trans-Mexican Volcanic Province, discussed in the next chapter.

Tectonic history

The western edge of the North American continent behaved as a passive margin throughout the Palaeozoic, accumulating large thicknesses of sediment deposited on the continental shelf and slope. Subduction of the Farallon Plate commenced in the early Jurassic, when several volcanic arcs and marginal basins were formed offshore. When the convergence pattern changed later in the Jurassic due to the opening of the Central Atlantic, the earlier-formed terranes were accreted to the continent, to be followed by further accretion during the early part of the Cretaceous. This accretionary zone formed the basement of the western half of the Cordillera.

In the late Cretaceous, a further change in the plate-tectonic regime caused an increase in convergence rate leading to a shallower subduction angle. This in turn gave rise to the thin-skinned Sevier fold-thrust belt, followed in the Eocene by the Laramide event, which resulted in the orogenic belt expanding further into the foreland. This period also saw the emplacement of the Sierra Nevada Batholith in the accretionary zone. During the Miocene, the over-thickened crust in the western part of the Cordillera experienced major extension and thinning, leading to the formation of the Basin and Range Province.

The most recent phase of volcanic activity commenced in the Eocene with the extrusion of the lavas of the Sierra Madre Occidental, followed by the voluminous Columbia River flows and the Cascade Arc. Activity continues at present in the Cascade Arc and at Yellowstone, at the eastern limit of the Columbia system.

Figure 11.10 Mount Whitney: the highest peak in the Sierra Nevada, carved from granite of the Sierra Nevada Batholith. © Shutterstock © Byron W. Moore.

12

Central America, the Andes and Antarctica

Central America and the Caribbean

There are two main elements of the late Mesozoic to Present-day orogenic system in the Central America–Caribbean region: the Central American orogenic belt, and the island chains of the Caribbean, which are divided into the Greater and Lesser Antilles (Fig. 12.1). The Central American belt joins southern North America to northern South America through the Panama Isthmus. The greater part of Mexico, including the Sierra Madre Occidental Volcanic Province, is part of the North American Cordillera and was described in the previous chapter. At the southern end of the Sierra Madre Occidental is the active Trans-Mexican Volcanic belt, which ends at the narrow Tehuantepec Isthmus. East of the isthmus is what is called here the 'Central American sector' of the belt, divided into four tectonic blocks: the Maya, Chortis, Chortuga and Choco Blocks, discussed below. The latter two blocks make up the Panama Isthmus, which is curved into a remarkable S-shaped bend, centred on the Panama Canal. The eastern end of the Choco Block is joined to the Colombian coast.

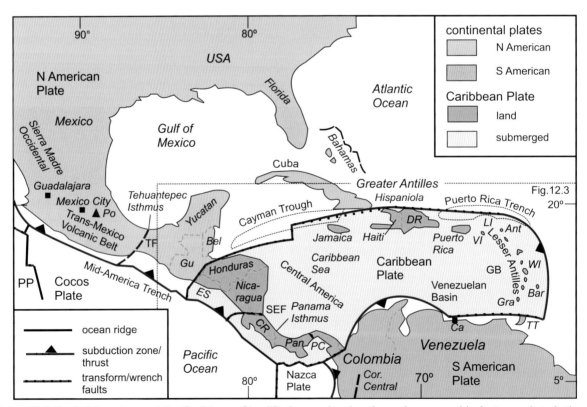

Figure 12.1 Central America and the Caribbean. Simplified map showing the main geographic features. Ant, Antigua; Bar, Barbados; Bel, Belize; Ca, Caracas; CR, Costa Rica; DR, Dominican Republic; ES, El Salvador; Gra, Grenada; Gu, Guatemala; LI, Leeward Islands; Pan, Panama; PC, Panama Canal; Po, Popocatepetl Volcano; TT, Trinidad & Tobago; WI, Windward Islands. Tectonic features: GB, Grenada Basin; PP, Pacific Plate; SEF, Santa Elena Fault; TF, Tepuantepec Fault. After James, 2013.

The Trans-Mexico Volcanic Belt

This volcanic belt, running through southern Mexico, lies at the southernmost end of the North American Cordillera, immediately south of the Sierra Madre Occidental. It is an active (Neogene–Recent) volcanic arc 1000km in length and 160,000km^2 in area, at its widest occupying the whole 'waist' of southern Mexico from the Pacific coast to the Gulf of Mexico. This area contains many active volcanoes over 4000m in height, including the famous Popocatepetl (Fig. 12.2), just south of Mexico City. The highest peak in the range is Pico de Orizaba (5636m) just east of Popocatepetl.

The Trans-Mexico Volcanic Belt results from the subduction of the Cocos Plate beneath the northern part of the Mid-America Trench, which follows the southern coast from just west of Guadalajara to the Tehuantepec Isthmus in Mexico. Interestingly, the arc containing the active volcanoes is oriented E–W – oblique to the trend of the trench. There is a short break in the arc at its intersection with the Tehuantepec Fault, after which it is replaced by the Central American Volcanic Arc.

The Central American sector

The circum-Pacific orogenic belt continues from southern Mexico through Guatemala, Honduras and Nicaragua, then follows the narrow, twisting Panama Isthmus to reach the northwestern coast of Colombia, where it joins the Cordillera Central at the northern end of the Andes. Although in politico-geographic terms 'Central America' excludes the country of Mexico, it is convenient to include the southernmost part of Mexico, known as the Maya Block, with the remainder of the Central American belt, which comprises the southwestern boundary of the Caribbean Plate, and is divided into three sections, the Chortis, Chortega and Choco Blocks (Fig. 12.1). The four blocks are separated by major fault zones.

The Maya Block

East of the Trans-Mexican Volcanic Belt, at the Tehuantepec Isthmus, there is a break in the mountain belt, which is displaced sinistrally along the N–S Tehuantepec strike-slip fault, to continue through Guatemala. The section of the belt from the isthmus up to the Honduras

Figure 12.2 Popocatepetl. Active volcano in the Trans-Mexico Volcanic Belt near Mexico City. Shutterstock©Kuryanovich, Tatsiana.

border is known as the Maya Block, and consists of a crystalline Precambrian and Palaeozoic basement overlain by Mesozoic sedimentary and volcanic strata deformed and metamorphosed in the Laramide Orogeny. This sector is similar to the accretionary complexes further north along the western margin of the North American Plate, and discussed in the previous chapter (see Fig. 11.7). There is no indication of a fundamental break between the Maya Block and the rest of Mexico. The foreland here is mostly concealed beneath the Gulf of Mexico, but the thick undeformed Mesozoic platform cover on the Yucatán Peninsula overlies continental-type crust.

The Chortis Block

Between Guatemala and Honduras, the belt is offset again by the major Motagua Transform Fault Zone, which marks the northwestern boundary of the Caribbean Plate (Fig. 12.3). The portion of the belt between here and Costa Rica is known as the Chortis Block. It

has been estimated that considerable sinistral strike-slip displacement has taken place along this fault zone. The Chortis Block is composed of a Mesozoic–Cenozoic sequence, similar to that of the Maya Block, underlain by a late Precambrian–Palaeozoic basement, and although its geology cannot be directly correlated with the Maya Block succession, it is considered that the two blocks are separated only by the strike-slip fault displacement rather than by a major suture.

The Chorotega Block

This block, consisting of Costa Rica and western Panama, is separated from the Chortis Block by the E–W Santa Elena Fault Zone along the Nicaragua–Costa Rica boundary, and from the neighbouring Choco Block by the Panama Canal Fault Zone. Both this block and the neighbouring Choco Block consist of an assemblage of oceanic and volcanic-arc material emplaced during the Eocene on a basement of Cretaceous volcanics. The nature of the pre-Cretaceous basement is unknown

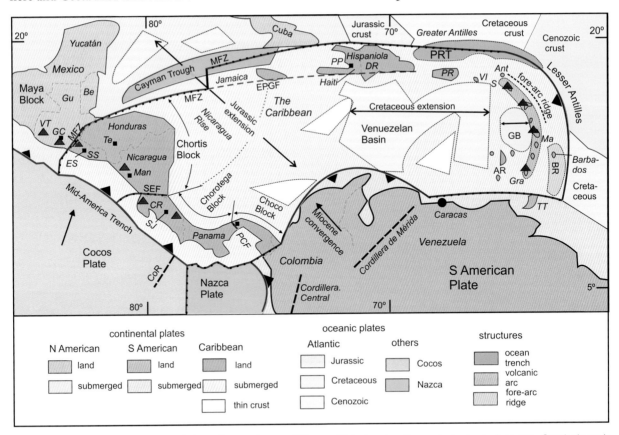

Figure 12.3 The Caribbean Plate. Enlarged central part of Figure 12.1 showing the main tectonic features. Symbols as in Figure 12.1, plus: AR, Aves Ridge; BR, Barbados Ridge; EPGF, Enriquillo–Plantain Garden Fault; MFZ, Motagua Fault Zone; PCF, Panama Canal Fault Zone. Geographic names: GC, Guatemala City; Ma, Martinique; Man, Managua; PP, Port au Prince; SJ, San José; SS, San Salvador; Te, Tegucigalpa; VI, Virgin Islands; VT, Volcán Tajumulco. Volcanoes: red triangles (only a few shown to indicate the extent of the volcanic belt). After James, 2013.

but the crustal thickness of over 40km on Costa Rica is indicative of continental-type crust.

The Choco Block

East of the Panama Canal Fault Zone, the orogenic belt describes an S-shaped bend before it attaches to mainland South America. This section of the Panama Isthmus is known as the Choco Block and consists of similar material to the Chorotega Block, although the structures are oriented differently. It is considered that the Chorotega–Choco sector has been transported northwards relative to both Chortis and South America by a combination of bending and NW–SE strike-slip faulting in response to the convergence between North and South America during the Miocene (see Fig. 4.4D).

There is no evidence that the North and South American continents were separated by oceanic crust anywhere along the isthmus now connecting the two, and it is thought that the original connecting crust was merely stretched, then subsequently bent, rather than broken.

The Central American Volcanic Arc

This active volcanic arc extends for a distance of about 1500km from Guatemala to northern Panama, and includes several hundred volcanoes, many of which are currently active (Fig. 12.3). The tallest is the dormant Volcán Tajumulco in Guatemala which, at 4220m, is the highest peak in Central America. The arc parallels the south coast of Central America and results from the subduction of the Cocos oceanic plate along the Mid-America Trench.

The most recent volcanic phase commenced in the mid-Miocene with the separation of the Cocos Plate from the old Farallon Plate. The western end of the arc occurs at the on-land break between the Trans-Mexican and Central American arcs, and corresponds to its intersection with the Tehuantepec Fault at the western end of the Maya Block. The Cocos Plate is currently subducting northeastwards beneath Central America at a rate varying from 50 to 80mm/a.

The Caribbean

The Caribbean region includes the two outer arcs of the Greater and Lesser Antilles, together with the Caribbean Sea enclosed between them and the mainland of Central and South America (see Fig. 12.1).

The Greater Antilles Arc

This island arc embraces the larger Caribbean islands of Cuba and Hispaniola (Haiti plus the Dominican Republic) and extends east to Puerto Rico and the Virgin Islands, a total distance of about 2500km. The arc is not now volcanically active, but was formed during the Jurassic and Cretaceous as a result of the northward subduction of the oceanic Caribbean Plate. Both Cuba and Hispaniola consist of basaltic and volcanic arc rocks thrust over Mesozoic platform deposits. These lie in turn on a late Precambrian–early Palaeozoic continental basement similar to that of Florida, and represent the southern extended part of the North American continent. The eastern part of the arc, from Hispaniola to the Virgin Islands, lies along the northern transform fault boundary of the active Caribbean Plate. This part of the arc was volcanically active into the Palaeogene, but activity ceased in the mid-Eocene.

The Caribbean Plate

The currently active Caribbean Plate (Fig. 12.3) is bounded on its northwestern side by the Motagua–Swan transform fault, which follows a gently curved path from the Motagua Fault Zone at the Guatemala–Honduras boundary through the Cayman Trough, passing between Cuba and Haiti, and along the northern coast of the Dominican Republic to join the western end of the Puerto Rico Trench. From there it continues eastwards until the trench bends south-eastwards, east of the Virgin Islands, where the plate boundary becomes the Lesser Antilles Subduction Zone.

The Lesser Antilles Arc meets the South American mainland near the island of Trinidad, where it joins a transform fault that runs parallel to the northern coast of Venezuela to just west of Caracas, where it joins an arcuate thrust that in turn connects with the eastern end of the Panama Isthmus.

The Caribbean Plate is currently expanding towards the Atlantic, assisted by the eastwardly-extending transform faults that form its north and south boundaries. The disastrous 2010 Haiti earthquake was caused by sinistral strike-slip movement on the Enriquillo–Plantain Garden Fault, which is a branch of the main transform fault extending E–W along the south side of Hispaniola.

The Lesser Antilles Arc

This arc comprises the island chains of the Leeward and Windward Islands (northern and southern sectors of the

Lesser Antilles, respectively) bordered on their outer side by the Puerto Rico Trench, and ending at Grenada near the Venezuelan coast. The currently active volcanic inner arc extends southwards from the small island of Saba, west of Antigua in the Leeward Islands, forming a wide arc facing eastwards towards the Atlantic Ocean. The trench is up to 9200km deep at its western end but becomes shallower eastwards and southwards, until by Barbados it has become completely filled by sediment and is replaced by an accretionary prism forming the outer fore-arc ridge (Fig. 12.4). Between the volcanic arc and the trench is an inactive outer arc consisting of several islands, including Antigua in the north and Barbados in the south, joined by a submerged ridge. There are 17 active volcanoes in the inner arc, including Mt. Pelée on Martinique (Fig. 12.5). Mt. Pelée is famous for its 1902 explosive eruption, which left about 30,000 dead, and gave rise to the concept of the 'Peléan' type of vulcanicity characterised by hot, glowing clouds of volcanic ash and catastrophic pyroclastic flows.

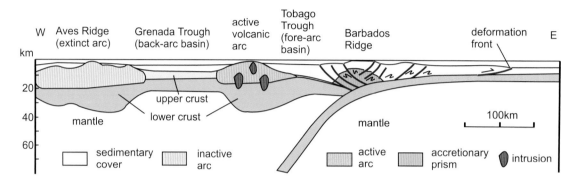

Figure 12.4 The Lesser Antilles Arc. Crustal section based on seismic data across the Barbados Ridge. Note that the deformation front has migrated eastwards some distance from the site of the original trench. After Westbrook, 1982.

Figure 12.5 Mt Pelée, Martinique – site of the disastrous 1902 explosive eruption that gave rise to the 'Peléan-type' of vulcanicity. Shutterstock©Albert Barr.

Tectonic History of the Central America–Caribbean region

The origins of the strange geometry of the Caribbean region can be traced back to the break-up of the Pangaea Supercontinent in the Jurassic, and are briefly summarised in Chapter 4 (see Fig. 4.4). Figure 4.4A shows that in the Triassic, South America was joined to North America but, as a result of the opening of the Central Atlantic during the Jurassic and early Cretaceous, North America rotated clockwise away from Africa, which was still joined to South America. This movement resulted in NW–SE extension between the Gulf Coast of the USA and northern South America, on the one hand, and a substantial sinistral strike-slip displacement on the other, as shown in Figure 4.4B. The amount of displacement should be roughly equivalent to the 1600km of Jurassic–early Cretaceous ocean crust formed off the North American coast up to the point when the Mid-Atlantic Ridge spread into the South Atlantic. Part of this displacement was taken up by sinistral movement along the Motagua–Swan Transform Fault, which might account for about half, and the remainder by the NW–SE extension of the region between the North and South American Plates.

The effect of the NW–SE extension was to cause thinning of the continental crust of the Gulf of Mexico and the Caribbean Sea, forming ocean basins in the centre of the Gulf of Mexico and east of the Yucatán Peninsula. The area between Central America and the Greater Antilles was also extended due to the retreat of the Lesser Antilles Arc into the Atlantic, and subsided to form the Caribbean Sea. The extension was uneven across this region, leaving elevated submarine areas such as the Nicaragua Rise.

A compressive phase occurred in the later Cretaceous and early Palaeogene due to a combination of westwards push from the Mid-Atlantic Ridge and an increasing convergence rate at the subduction zone along the Pacific margin. This gave rise to the widespread obduction of oceanic material, including major ophiolite bodies, in Central America and the Greater Antilles Arc.

From the late Eocene until the Present, the Caribbean Plate expanded and grew eastwards due to the retreat of the Lesser Antilles Subduction Zone as the Atlantic crust subducted beneath it. An additional 300km was added to the Caribbean Plate as a result.

In the Miocene, northwards convergence of South America with Central and North America resulted in bending and north-eastwards displacement of the Panama Isthmus and northernmost South America.

The Andes

The Andes mountain belt follows the western boundary of the South American continent for 7600km from northern Colombia to Tierra del Fuego in southern Chile, spanning 65° of latitude from 10°N to 55°S (Figs. 12.6, 12.7). For most of this distance, the belt is a rather narrow mountain spine only about 200km wide, but it broadens to over 700km width in the central section through Bolivia. Only the northernmost part deviates from this simple pattern, dividing into three branches as it passes through Colombia. The Andes are the longest, and apart from the Himalayas, the highest on-land mountain belt on Earth. They traverse seven countries, from north to south: Venezuela, Colombia, Ecuador, Peru, Bolivia, Chile and Argentina; for most of their length they divide the narrow coastal belt from the remainder of the continent, and there are few routes across.

Figure 12.6 The Andes and the Columbian chains. CC, Cordillera Central; CM, Cordillera Mérida; SP, Sierra de Perija. Shutterstock©Arid Ocean.

The orogenic belt can be divided into three sectors: northern, extending from the north coast of Colombia and Venezuela to southern Ecuador, which trends NNE–SSW to N–S; central, trending NW–SE through Peru and Bolivia; and southern, trending N–S, through Chile and Argentina, but bending round almost to E–W in Tierra del Fuego. The junction between the central and southern sectors in Bolivia, which involves a change in orientation of 35°, is known as the Arica Embayment or 'Arica elbow'. This area corresponds to the broadest part of the belt, containing the elevated plateau of the Altiplano, much of which is over 4000m in height and contains several of the highest peaks in the Andes.

Plate-tectonic summary

In plate-tectonic terms, the Andean belt presents a relatively simple scenario: an active plate boundary is defined by a subduction zone along which the Nazca and Antarctic oceanic plates are descending eastwards beneath the South American continent along the Peru–Chile Trench, and an active volcanic arc follows the mountain belt for most of its length. It has long been regarded as the type example of an active continental margin – the 'Andean margin', where the deformation and uplift of the orogenic belt are due to the subduction process alone, and continent–continent collision has played no part.

The volcanic arc is not continuous throughout; there is a gap between the Ecuador–Peru border and the point where the Nazca Ridge meets the Peruvian coast about 150km southeast of Lima. This sector is known as the Peruvian 'flat-slab' segment, attributed to a subduction angle too shallow to intersect the zone of magma generation and a relatively fast convergence rate of 70–90mm/a. The distribution of seismicity indicates a shallow Benioff Zone here, whereas to the north and south, it indicates slab dips of over 30°. A second flat-slab sector is present in the Chilean sector from south of Antofagasta to where the Juan Fernandez Ridge intersects the coast at 33°S, near Valparaiso. South of the Chile triple junction, where the Chile Rise meets the subduction zone, the Antarctic Plate is subducting much more slowly, about 20mm/a beneath Patagonia.

From the late Cambrian Period, through the remainder of the Palaeozoic, and during the early Mesozoic, the western border of what is now South America was an active continental margin, where oceanic crust of the ancestral Pacific was being consumed. The tectonic history of the present orogenic belt began with extensional rifting in the Triassic, as the long process of break-up of Pangaea commenced. During the Jurassic, volcanic arcs developed offshore, with back-arc basins on their landward side. With the opening of the South Atlantic Ocean during the Cretaceous, changes in the subduction regime along the Pacific margin prompted the commencement of compression and uplift along what became the Andes belt. Subduction continued with its attendant vulcanicity through the Cenozoic as the orogenic belt grew, and several large foreland basins, such as the Orinoco and Amazon Basins, developed on the South American craton to the east. The present structure had taken shape by the mid-Miocene.

The Northern Sector – Venezuela to Ecuador

This sector includes the part of the Andes belt that passes through Ecuador and Colombia, ending at the northern coast of Venezuela (Figs. 12.7, 12.8A). The mountain belt divides after it crosses the Colombian border, one branch, the Cordillera Central, proceeding in a NNE direction and dying out west of Bucaramanga to be replaced by a wide coastal plain. This part of the belt includes many peaks over 5000m in height and numerous active volcanoes. The second branch, known as the Cordillera Oriental, equally high but lacking active vulcanicity, strikes north-eastwards and crosses the Venezuelan border to reach the coast west of Caracas. The Venezuelan sector of this chain is named the Cordillera Mérida. At the City of Bucaramanga, another mountain chain, the Sierra de Perijá, branches off and takes a sinuous course northwards towards the coast. The Central and Eastern Cordillera are separated by a broad valley containing the capital city of Bogotá.

In tectonic terms, the northern sector can be divided into a western accretionary zone consisting mainly of oceanic rocks, a narrow central zone dominated by Cenozoic volcanic rocks that overlie the Precambrian basement, together with numerous plutonic intrusions, and an eastern zone corresponding to the foreland fold-thrust belt (Fig. 12.8A). The western accretionary complex is separated for much of its length from the central zone by a major dextral strike-slip fault complex – the Dolores–Guayaquil Megashear. Much of the Western Zone is below sea-level, but the on-land part includes several terranes consisting of thrust packages of oceanic material.

In the central zone, the Precambrian basement with its plutonic intrusions has been uplifted to form the high mountains of the Cordillera Central. The eastern zone

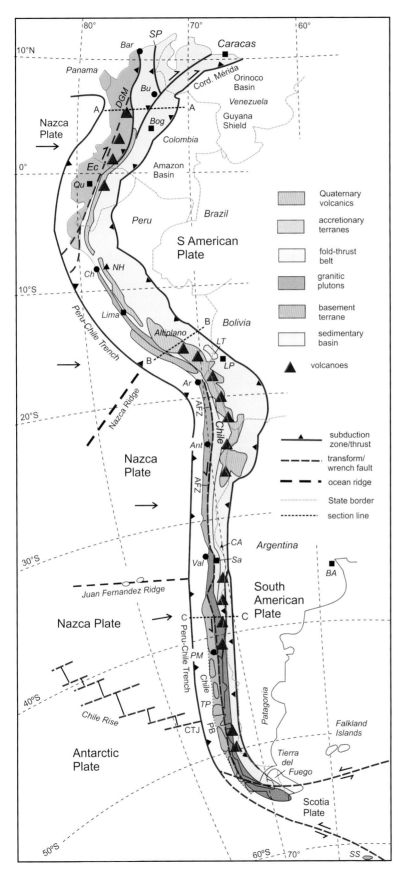

Figure 12.7 The Andes. AFZ, Atacama Fault Zone; Ant, Antofagasta; Ar, Arica; BA, Buenos Aires; Bar, Baranquilla; Bog, Bogota; Bu, Bucaramanga; CA, Cerro Aconcagua; Ch, Chimbote; CTJ, Chile Triple Junction; DGM, Dolores-Garcia Mega-shear; Ec, Ecuador; LP, La Paz; LT, Lake Titicaca; NH, Navado Huascarán; PB, Patagonian Batholith; PM, Porto Montt; Sa, Santiago; SP, Sierra de Perija; SS, South Shetland Islands; TP, Taitao Peninsula; Val, Valparaiso. Note that volcano symbols indicate groups of individual volcanoes. After Moores & Twiss, 1995.

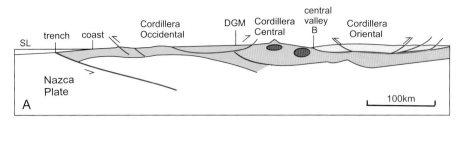

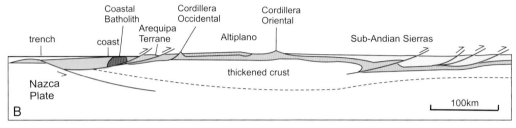

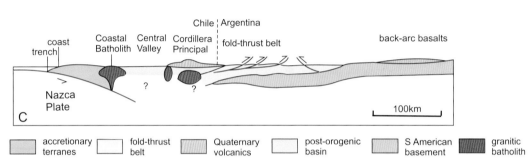

Figure 12.8 Schematic sections across the Andes. **A**. Northern sector: Colombia to Venezuela, taken along E–W line A–A* through Bogota. **B**, Central sector: SW–NE line B–B through Peru, from just north of Arica. **C**, Southern sector: E–W line C–C from south of Valparaiso. B, Bogota; DGM, Dolores-Garcia megashear. Note that section B represents the Peruvian 'flat-slab' segment and lacks contemporary vulcanicity. [*See lines of section on Figure 12.7.] After Moores & Twiss, 1995.

contains a thick sequence of Palaeozoic to Cenozoic rocks deposited on the Precambrian basement of the South American Plate. This zone broadens out northwards to over 600km near Bucaramanga, at which point the Cordillera Oriental divides into the Sierra de Perijá and the Sierra de Mérida. The elevated parts of the eastern zone, including the Cordillera Oriental and its northern branches, correspond to the uplifted sectors of the fold-thrust belt and are separated in the north by wide unfolded regions. These belts have all experienced dextral transpressional movements as a result of the oblique (E–W) convergence direction of the Nazca Plate. The Venezuelan part of the fold-thrust belt is a response to the Miocene convergence between South and North America.

East of the fold-thrust belt is the South American foreland, here represented by the Guyana Shield, which is unaffected by Andean deformation.

The Central Sector – Peru and Bolivia

As the Andes pass from southern Ecuador into northern Peru, both the mountain belt and the coast bend abruptly through 60° to trend NW–SE, oblique to the E–W convergence direction of the Nazca Plate, but in the opposite sense to that of the northern sector, which accounts for the presence of sinistral transpressive structures in the coastal zone. The volcanic arc of the northern sector ends near the Ecuador border and only reappears 500km to the south, near the Bolivian border. This volcano-free sector is the 'Peruvian flat-slab segment' referred to above, and is attributed to the faster convergence rate of the Nazca Plate. Here the western accretionary zone is very narrow and almost entirely submerged (Fig.12.8B). The belt of coastal mountains, the Cordillera Occidental, is composed largely of a compound batholith of late Jurassic to Cretaceous age that has been intruded into Precambrian basement of the

Arequipa Terrane. This allochthonous terrane is thought to have originally formed part of the South American foreland, but was thrust eastwards over the Mesozoic cover of the central zone during the Andean Orogeny.

The remainder of the Peruvian sector belongs to the very broad foreland fold-thrust belt, which is composed of a Mesozoic to Cenozoic sedimentary cover overlying the Precambrian basement of the foreland and deformed in the Andean Orogeny. Near the coastal batholith, the cover consists of oceanic-type rocks but grades inland into strata representative of the continental slope and platform. The sedimentary cover is overlain by Quaternary volcanics; these form a narrow belt in the north but broaden out near the Bolivian border.

The western part of the fold-thrust belt has been elevated to form the Cordillera Occidental and was deformed during the Eocene, whereas the central part is relatively flat-lying and forms the high plateau of the Peruvian Altiplano. East of the plateau, in the Cordillera Oriental, the basement has been uplifted during a later phase of the Andean Orogeny and affected by steep reverse faults and open folds. The Sub-Andean Zone, east of the Cordillera Oriental, is characterised by imbricate NE-directed thrusting of Neogene age.

Both the western and eastern Cordillera contain peaks up to over 6000m high, and the Altiplano Plateau itself is around 4000m high. The highest peak in Peru, at 6768m, is Nevado Huascarán (Fig. 12.9) in the Cordillera Occidental, east of Chimbote. Somewhat confusingly, there are three parallel ranges in the north: western, central and eastern cordillera, separated only by narrow valleys, whereas in the south the central plateau separating the western and eastern ranges is 200km across, and is occupied by the famous Lake Titicaca. This high plateau broadens further to about 400km wide in northern Bolivia, where it is known as the Bolivian Altiplano. This sector corresponds to the 'flat-slab' segment of the subduction zone and mostly lacks vulcanicity. Here the crust is up to 70km thick, the whole region is elevated to above 4000m, and the central valley has disappeared.

The amount of shortening across the Peruvian fold-thrust belt has been estimated at 120–150km – around 25%. This figure is consistent with the amount of westward movement of South America (relative to a fixed trench position) during the Neogene.

The Southern Sector – Chile and Argentina

The junction between the Central and Southern sectors of the Andean belt is marked by another abrupt change in orientation of both the coastline and the mountain range from NW–SE to N–S, and occurs at the Arica embayment near the Peru–Chile border (Fig. 12.7). The Chilean Andes are over 4000km long and span about 37° of latitude. The range is relatively narrow, for the most part under 200km in width, but it broadens in Northwest Argentina and Bolivia to about 550km. In the north, there is a very steep rise in elevation from

Figure 12.9 Nevado Huascarán, 6768m. Extinct volcano and highest peak in the Peruvian Andes. Shutterstock©Christian Vinces.

the coastal plain, here occupied by the Atacama Desert, to over 6000m. In much of the southern sector (Fig. 12.7), the coastal plain is almost non-existent, but there are numerous islands forming a chain parallel to the coast. The highest peak in the southern sector is Cerro Aconcagua (6962m) 150km east of Valparaiso, and just across the border with Argentina.

As in the Peruvian sector, the belt is accompanied offshore by the Peru–Chile Trench, which attains its greatest depth of over 8000m just south of the Arica embayment. From its intersection with the Juan Fernandez Ridge southwards, the trench is filled with sediment, and consequently not obvious as a topographic feature. South of the Chile Triple Junction, where the trench meets the Chile Rise, it is replaced by a transform fault. From here southwards, the Andes are flanked by the Antarctic Plate.

There are again three main tectonic zones: western, central and eastern. The western zone contains the Coastal Cordillera, which consists mostly of late Palaeozoic to Mesozoic igneous and metamorphic rocks, including the large Patagonian Coastal Batholith of Permian age. South of Puerto Montt, this zone consists of the many islands and archipelagos that fringe the west coast. Much of this southern sector consists of a deformed and metamorphosed accretionary prism of Mesozoic age. The eastern margin of the western zone is defined over much of its length by the Atacama Fault Zone, which runs parallel to the coast for over 1000km from the Peruvian border to near Valparaiso. Another fault zone forms the boundary between the coastal zone and the Patagonian Batholith in the southern part of the sector. These fault zones exhibit sinistral transpressional characteristics and are believed to date back to the Mesozoic. The more recent structures are a response to E–W compression, reflecting the present E–W convergence direction.

The central zone consists of a low-lying area formed by a Mesozoic to Cenozoic sedimentary basin; between Santiago and Puerto Montt, this region provides most of Chile's agricultural land and hosts several of its cities. South of Puerto Montt, the central zone is absent.

The eastern zone contains the main Andean mountain range, the Cordillera Principal, and corresponds to the fold-thrust belt, which was uplifted during the Miocene, and is still actively rising. The height of the mountain range decreases southwards towards Tierra del Fuego from over 6000m in Northern Chile to under 4000m in Patagonia.

Volcanoes form an almost continuous line 200–400km from the coast, from the Peru–Chile border to the Chile Rise, generally along the Chile–Argentina border. There is one gap, between 27°S and 31°S, attributed to the second flat-slab section of the subduction zone referred to earlier.

Near the southern end of the belt, the Peru–Chile Trench is intersected by the Chile Rise, which separates the Nazca and Antarctic Plates. This ridge currently lies offshore from the Taitao Peninsula but has been slowly migrating northwards at a rate of about 20mm/a as the Antarctic Plate expands. South of the Chile Rise, the offshore plate boundary, now separating the South American from the Antarctic plates, becomes a transform fault. The southernmost sector of the belt, through southern Patagonia and Tierra del Fuego, bends round into an NW–SE orientation. Here it is influenced by a zone of sinistral strike-slip faults that define the northern boundary of the Scotia Plate.

Antarctica and the Scotia Arc

The former continuation of the Andes belt into the Antarctic Peninsula was broken during the middle Jurassic when Gondwana split up and East Gondwana (Antarctica–Australia–India) moved away from South America–Africa. As South America began to move westwards relative to Antarctica in the mid-Cretaceous, the land bridge between the two continents became extended, then split into several fragments, including South Georgia and the South Orkney Islands. By the early Eocene, the increasing separation between South America and Antarctica had created the Scotia Sea, which was bounded on its north and south sides respectively by the North and South Scotia Ridges.

The Scotia Sea Plate

This small plate (Fig. 12.10) is defined on its northern and southern sides by transform fault zones along the North and South Scotia Ridges respectively, both of which are northwardly convex arcs. The western side is formed by the Shackleton Fracture Zone, which links with the end of the Peru–Chile Trench to the north. The eastern side of the plate is defined by the active volcanic Scotia Arc. The most easterly section of the Scotia Sea Plate is the most active part and forms a microplate, the South Sandwich Plate, separated from the main plate by a spreading ridge. This microplate was formed as a result of back-arc expansion from the South Sandwich Subduction Zone during the last eight million years.

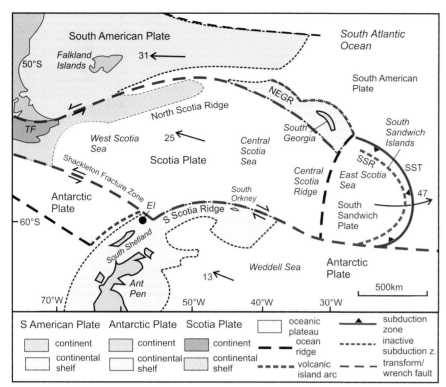

Figure 12.10 The Scotia Plate. The active Scotia volcanic arc forms the eastern boundary of the Scotia Plate; the northern and southern boundaries are transform faults along the North and South Scotia Ridges respectively; the western boundary is also a transform fault that links with the Peru–Chile subduction Zone. EI, Elephant Island; NEGR, NE Georgia Rise; SSR, South Sandwich Ridge; SST, South Scotia Trench. After joannenova. com.au., via Wikimedia Commons.

The Scotia Arc

There are eleven volcanoes on the small islands and seamounts in this 400km-long arc, most of which are currently active. On the Atlantic side of the arc is the South Sandwich Trench, along which Atlantic oceanic crust belonging to the South American Plate is subducting at the relatively fast rate of 70–90mm/a. At its southern end, the arc meets the transform fault along the South Scotia Ridge. From this arc–fault junction, the South American plate boundary strikes eastwards into the Atlantic to link up with the Mid-Atlantic Ridge.

Antarctica

Although much of the Antarctic continent is still largely unknown, it is clear that the whole of it now lies within the Antarctic Plate, whose northern boundary runs from the Chile Rise, along the Shackleton Fracture Zone to Elephant Island, and from there follows the South Scotia Transform Fault towards the American–Antarctic Ridge, as shown on Figure 12.10. Other ridges surround the continent on its eastern, southern and western sides. However, during the early Mesozoic, Western Antarctica was the active margin of this part of Gondwana, and a subduction zone with its accompanying volcanic arc extended from the southern end of the Andes along the Antarctic Peninsula and the

West Antarctic Mountains (Fig. 12.11). East Antarctica, which comprises the larger part of the continent, consists of a late Precambrian craton with a younger platform cover. The craton is bounded on its western side by the Transantarctic Mountains, which represent the uplifted western margin. During the early Cenozoic, an extensional rift running from the Weddell Sea to the Ross Sea developed between West and East Antarctica, and is still tectonically active.

There are about 37 known volcanoes in Antarctica, at least five of which are currently active. These include Mount Melbourne and the famous Mount Erebus, which lie along the southern part of the Trans-Antarctic Mountains. Erebus, situated on Ross Island, was first climbed by members of Ernest Shackleton's first polar exploration party in 1908. It is thought that the vulcanicity of this belt is related to the extensional rift, and it has been suggested that it may originate from a mantle plume rather than from a currently active plate boundary.

There are still two active volcanoes on the Antarctic Peninsula, opposite the South Shetland Islands, together with at least a further two that are extinct or dormant. These are related to a short stretch of the now inactive subduction zone that lay northwest of the South Shetland Islands. This subduction zone meets the Shackleton Fracture Zone and the South Scotia Transform Fault

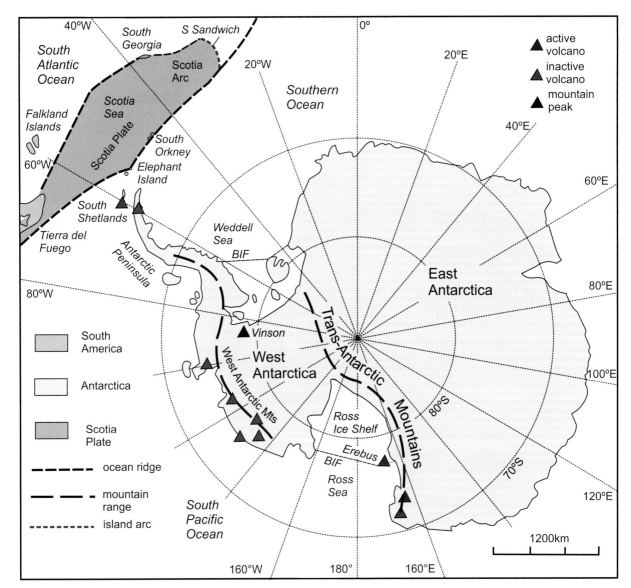

Figure 12.11 Antarctica. The Antarctic Peninsula and the West Antarctic Mountains represent a Mesozoic volcanic arc that was added to the Palaeozoic East Antarctic Craton. The Trans-Antarctic Mountains are the uplifted margin of the old craton. The region between the two mountain ranges, extending from the Weddell Sea to the Ross Sea, is an extensional rift. The Vinson Massif, at 4897m, is the highest peak. Only a representative selection of the extinct volcanoes is shown. BIF, Barrier ice front. After Bulkeley, 2008.

at a triple junction near the tiny Elephant Island, at the northeast end of the South Sandwich Island chain (see Fig. 12.10).

It was on Elephant Island that Sir Ernest Shackleton and his 1914 expedition spent the winter of 1915 after their ship was crushed by the sea ice. The subsequent 1300km journey across the Scotia Sea undertaken by Shackleton and 5 companions in an open boat to South Georgia is one of the most remarkable exploits in the history of polar exploration.

13

The ocean ridges

Were the world's oceans to be removed, a great system of mountain ranges, exceeding even the Himalayas in scale, would be revealed (Fig. 13.1). This mostly submerged system, extending for a total length of 65,000km throughout the oceans, was largely unknown until the introduction of electronic echo-sounding in 1923. Prior to this development, only sporadic soundings by conventional lead-line methods had been available.

Historical investigations

The presence of a ridge in the mid-Atlantic was well known by the late nineteenth century, and had been commented on by some of the early geologists, including Alfred Wegener, who incorporated it into his ideas of continental drift in 1912. The introduction of electronic echo-sounding enabled detailed surveys of the Mid-Atlantic and Carlsberg ridges to be completed by 1933. From the 1950s onwards,

a comprehensive programme of oceanic surveying was undertaken by scientists from several oceanographic institutions using more advanced techniques including gravity, seismic, heat-flow and magnetic surveys, together with side-scan sonar imaging, to obtain a detailed picture of the ridge topography. In 1977, Bruce Heezen and Marie Tharp from Lamont-Doherty Observatory in Columbia University published a topographic map of the whole mid-ocean ridge system (Heezen and Tharp, 1977). Subsequent investigations have included visual imaging by towed cameras, manned submersibles and remote-controlled vehicles. Further information came from a series of international ocean-drilling programmes – the Deep-Sea Drilling Project (DSDP, 1968–83) and its successor the Ocean Drilling Program (ODP, 1985–2004), which provided rock samples from the various parts of the ridges.

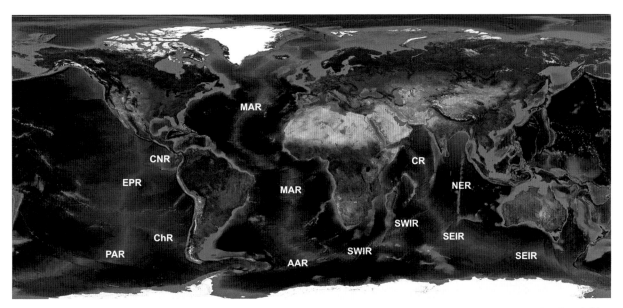

Figure 13.1 The ocean ridge network from space. Oceanic raised features are shown in paler blue. Note that in addition to the mid-ocean ridges, which are typically broad and often poorly defined, there are many more sharply defined ridges such as the very obvious Ninety-east Ridge and the many island arcs. Ridges: AAR, Atlantic–Antarctic; ChR, Chile; CNR, Cocos–Nazca; CR, Carlsberg; EPR, East Pacific; MAR, Mid-Atlantic; NER, Ninety-east; PAR, Pacific-Antarctic; SEIR, SE Indian; SWIR, SW Indian; Compare with Figure 13.2. © Shutterstock, by McLek, from NASA image.

Plate-tectonic background

The role played by the mid-ocean ridges in plate-tectonic theory was described in chapter 3 (see Figs 3.7 and 3.8). To recap briefly, they are the sites where new oceanic crust is generated by the upwelling of mantle-derived melts, and are referred to as divergent (or constructive) plate boundaries, or as ocean-spreading centres. It is important to note that oceanic ridges are also formed along transform faults and submerged volcanic arcs, and many examples of these were described in the previous chapters; only those that are volcanically active spreading centres form part of the mid-ocean ridge network as shown in Figure 13.2.

The controlling mechanism of the spreading ridge is ultimately governed by the heat distribution in the uppermost mantle; warmer material underlying the ridges and cooler material at the trenches produce the downward force of slab pull at the trenches and the extensional force of ridge push at the ridges, and it is these forces that provide the motor that drives the movement of the plates. It is the extensional state of stress within the ridge that allows magma to access the crust and not, as might have been thought, the pressure of the rising magma. The ridge-push force is caused by the outwards gravitational pressure exerted by the excess mass of the ridge topography – the same mechanism that was discussed in the context of on-land mountain ranges such as the Himalayas, where the effects of the rising orogenic edifice are countered by gravitational gliding on the mountain flanks.

Ridges do not, in general, lie above uprising mantle columns, and their topography is not directly related to the mantle convection pattern, but is governed by the stress state in the strong upper crust. Ductile asthenospheric mantle is drawn upwards into an extensional region, and melting occurs because of the consequent release of pressure. The initiation of spreading systems, however, does seem to take place above mantle hotspots or plumes, often creating 'triple junctions' from which rifts spread outwards, as in the case of the Red Sea–Gulf of Aden–African Rift described in chapter 3 (see Fig. 3.9). Here the continental crust has first been deformed into a dome above the plume, which subsequently cracks into three rifts under the extensional stress.

As oceanic plates grow and their geometry changes, the initial trend of the ridge axis may become oblique to

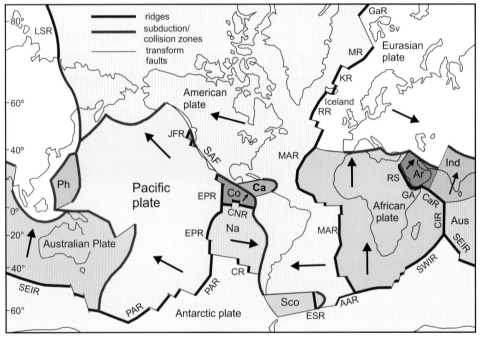

Figure 13.2 The ocean ridge network. AAR, American–Antarctic Ridge; CaR, Carlsberg Ridge; CIR, Central Indian Ridge; CNR, Cocos–Nazca Ridge; CR, Chile Rise; EPR, East Pacific Rise; ESR, East Scotia Ridge; GA, Gulf of Aden Rift; GaR, Gakkel Ridge; JFR, Juan de Fuca Ridge; KR, Kolbeinsey Ridge; LSR, Laptev Sea Rift; MAR, Mid-Atlantic Ridge; MR, Mohns Ridge; PAR, Pacific–Antarctic Ridge; RR, Reykjanes Ridge; RS, Red Sea Rift; SEIR, SE Indian Ridge; Sv, Svalbard; SWIR, SW Indian Ridge. Plates: Ar, Arabian; Aus, Australian; Ca, Caribbean; Co, Cocos; Ind, Indian, Na, Nazca, Ph Philippines; Sco, Scotia. The arrows give the direction of motion of each plate relative to the Antarctic Plate (regarded as stationary). After Vine & Hess, 1970.

the spreading direction. In this case, the ridge generally adjusts by forming short transform offsets. Lengthy sections of ridge that are oblique to the spreading direction are uncommon (one example is the Reykjanes ridge south of Iceland, which is discussed below).

Topography and structure of spreading ridges

In general terms, the spreading ridges are submarine mountain chains that can be between 1000km and as much as 4000km in width, and are elevated by between 1km and 2km from the surrounding ocean floor (Fig. 13.3). They typically incorporate a central rift valley, up to around 100km in width, along which the vulcanicity and associated seismic activity is concentrated. Individual ridges differ considerably in profile, depending on their spreading rate; faster-spreading ridges display a smoother topography with narrow, sharply-defined central rifts, whereas slower-spreading ridges are characterised by more rugged relief and wider central rifts (Fig. 13.4). The rifts are bounded by normal faults with displacements

of several hundred metres. Central rifts are absent from very fast-spreading ridges.

Spreading rates of individual ridges are described as slower in the range 10–55mm/a, medium, 55–100mm/a, and faster over 100mm/a (note that 1mm/a=1km/Ma). It is often convenient to quote the 'half-spreading' rate, which is the velocity of an oceanic plate relative to its spreading axis, although the half-spreading rates on each side are not necessarily the same. Spreading rates on most ridges, including the Mid-Atlantic and Indian Ocean ridges, are slow, whereas those of the East Pacific Rise are much faster.

The shape of the ocean ridges, like that of the on-land mountains, is governed ultimately by gravitational forces; because ridges are the sites of upwelling warm (and in part melted) mantle material, they are less dense than the surrounding cooler ocean floor, and the crust therefore expands to form a topographic bulge in order to attain gravitational equilibrium. However, compared with continental mountains, erosion plays a much reduced role in modifying their topography, which is governed essentially by a series of escarpments formed as a result

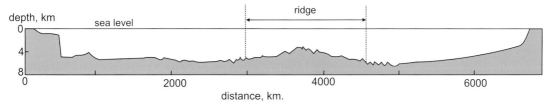

Figure 13.3 The Mid-Atlantic Ridge. Traverse from Eastern USA (Florida) to West Africa. The ridge is just over 1500km wide and occupies about one-third of the width of the Atlantic Ocean here, rising from a depth of about 6km to about 3km at the apex. After Searle, 2015.

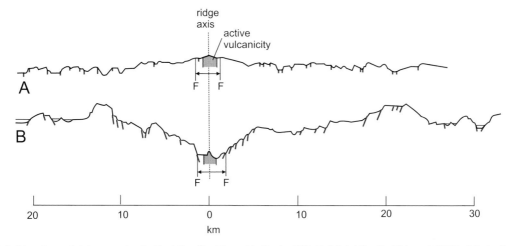

Figure 13.4 Structure of ridge crests. **A**, East Pacific Rise at latitude 3°S; **B**, Mid-Atlantic Ridge at 37°N. A is fast-spreading (64mm/a) and B slow-spreading (23mm/a), F–F, zone of active fissures; faults in red; zone of recent volcanic eruptions in purple; sedimentary basins in yellow. The fast-spreading ridges have a much smoother topographic profile. After Searle, 2015.

of the extensional faulting, or by volcanic edifices created by escaping lava.

As the new crust created at the spreading centre cools, it moves sideways to accommodate fresh injections of new material at the ridge axis, and ultimately sinks to the general level of the deep-ocean floor. The time taken, and thus the distance achieved, by this process depends on the spreading rate, which explains why faster-spreading ridges are broader and *vice versa*.

Temperature and heat flow variation

Because the ocean ridges are in a state of gravitational equilibrium, the extra surface mass must be compensated by less dense underlying material. Ridges have a characteristically cuspate topographic profile in which the slope becomes shallower with distance from the ridge axis. The height varies inversely with distance, and thus with the age of the underlying crust. This model conforms with the pattern of changing heat flow (Fig. 13.5), which varies from several hundred mWm^{-2} (milliwatts per square metre) at the ridge axis to about $50mWm^{-2}$ on the flanks, indicating that the ridge topography matches the temperature distribution, which in turn determines the density of the underlying lithosphere.

Another way of illustrating this pattern is by the variation in the thickness of the lithosphere, the base of which is determined essentially by the temperature at which its properties change from brittle to ductile, usually taken to be the 750° isotherm. Thus defined, the lithosphere thickness varies from zero at the ridge axis to 9.5km in 1Ma-old crust, 19km in 4Ma-old crust, and 95km in 100Ma crust. The breadth of the ridge is determined by the cooling rate, which depends in turn on the spreading rate; for fast-spreading ridges like the East Pacific Rise, this can be as much as 4000km, whereas in slow-spreading ridges such as the Mid-Atlantic, the width is less than 1000km for most of its length.

Earthquakes

It was the distribution of earthquakes along the mid-ocean ridges that helped to establish the concept of plates, whose boundaries were defined by the lines along which current tectonic activity appeared to be concentrated (see chapter 3, Fig. 3.1). These earthquakes occur within what is termed the seismogenic zone, within which the crust is susceptible to brittle deformation, and this zone lies at progressively deeper levels as the crust becomes older and cooler. Earthquake foci on the spreading ridges are usually confined to the central axial zones, and are invariably shallow: on fast-spreading ridges, they typically occur at depths of less than 1km, whereas on slow-spreading ridges, they may occur at depths of 7–10km. However in old, cold lithosphere, earthquakes can occur at much greater depths.

Earthquakes also occur along, and help to define, ridge offsets; these are the oblique or transverse sections, some of which correspond with transform faults that link adjoining segments of spreading ridge. Most earthquakes appear to be generated by extensional faulting in a tectonically active zone within a few tens of kilometres of the axis of spreading ridges, or by

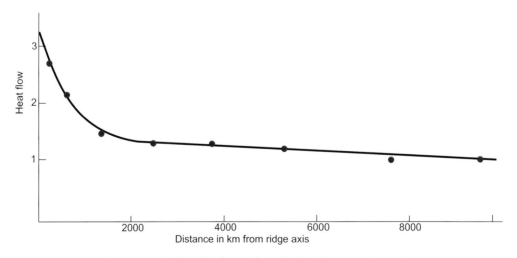

Figure 13.5 Variation in heat flow across the Pacific Ocean floor. Red circles are observed data. Note that the heat flow declines steeply from the axis of the East Pacific Rise, then decreases gradually with increasing age across the ocean floor. Heat flow data in microcalories per cm² per sec. After Uyeda, 1978.

strike-slip displacements on the transform faults. Some transform fault zones have experienced compressional or extensional movements due to subsequent changes in plate geometry, and these produce uplifted topographic ridges or depressed trenches, respectively.

Vulcanicity

Active vulcanicity is concentrated in the axial region of the mid-ocean ridges (Fig. 13.3); in slow-spreading ridges this takes place in the central rift valley, often forming an elongate ridge a few kilometres wide above the currently active rift that marks the plate boundary. Eruptions on fast-spreading ridges typically create rather shallower ridges a few kilometres in height, which extend along the whole length of the actively spreading sector. Axial ridges generally exhibit a narrow valley along their crest, a few kilometres wide and several hundred metres high, situated above the active fissure; this is caused by gravitational collapse when the lava supply becomes exhausted.

About 20 eruptions occur somewhere along the ridge network every year – more frequently on the fast-spreading ridges. Individual lava flows vary from pillow, tube or ropy forms similar to those seen in near-shore underwater eruptions, to sheet-like shapes, depending on the local topography. Pillow shapes predominate on the slower-spreading ridges with steeper slopes, whereas sheet flows are more typical of the faster-spreading examples with higher flow rates.

Mid-ocean ridge magmas are invariably basaltic, produced by the melting of dry lherzolitic peridotite from the upper mantle, and either extruded as lavas or intruded into the crust as gabbroic sheets. Hydrothermal activity is common, particularly on fast-spreading ridges, and results in the formation of pipe-like vents that form cones or pillars rising from the sea floor. These are termed 'black smokers', and are a rich source of ore mineralisation because of the interaction between hot basalt and saline hydrothermal fluid. Magmas erupted from the volcanic islands – the 'hotspot magmas' – are typically more alkaline in composition, thought to indicate their derivation from a more enriched, higher-temperature mantle source.

There is insignificant sedimentary cover anywhere near the ridge axis owing to the slow rate of deposition in the deep-ocean environment: only fine siliceous or calcareous ooze mixed with small quantities of very fine clay is deposited there at rates of only a few millimetres per million years. However, significant accumulations of sediment may occur on the flanks, particularly of slow-spreading ridges, where the crust is over 10Ma old, as seen in Figure 13.4.

The mid-ocean ridge network

The active mid-ocean ridge network spans all the world's oceans (see Figs 13.1, 13.2). The longest, and one of the best-studied, is the Mid-Atlantic Ridge. Major ridges also traverse the Pacific, Antarctic and Indian Oceans; the East Pacific Rise lies on the eastern side of the Pacific and continues south-westwards as the Pacific–Antarctic Ridge, which separates the Pacific Plate from Antarctica. Four separate ridge sectors are recognised in the Indian Ocean: the SW and SE Indian Ridges, the Central Indian Ridge and the Carlsberg Ridge. The short American–Antarctic Ridge separates the American Plate from the Antarctic Plate.

Short stretches of ridge occur on the margins of the smaller oceanic plates: the Scotia Ridge on the Scotia Plate; the Gulf of Aden Ridge (or Rift) along the Gulf of Aden, forming the south-western boundary of the Arabian Plate; the Chile Rise, between the Nazca and Antarctic plates (see Fig. 12.7); the Cocos–Nazca Ridge between the Cocos and Nazca Plates; and the Juan de Fuca Ridge between the Pacific and Juan de Fuca plates (see also Fig. 11.2). The northern part of the Mid-Atlantic Ridge itself is subdivided into a further four separate segments: the Reykjanes, Kolbeinsey, Mohns and Gakkel ridges. Some of the best-studied stretches of ridge are on the Mid-Atlantic, East Pacific, Nazca and Carlsberg ridges. The only place where the structures of the ocean ridge can be studied in detail on land is the island of Iceland, situated on the northern part of the Mid-Atlantic Ridge, where its unusual elevation is ascribed to a mantle hotspot located beneath the ridge.

The Atlantic
The Mid-Atlantic Ridge (MAR)

This is the longest of the mid-ocean ridges, extending for about 16,000km from its junction with the SW Indian Ridge in the Southern Ocean to the Siberian shores of the Arctic Ocean (Figs. 13.2, 13.6). It was the first of the mid-ocean ridges to be discovered, in the eighteenth century, and the first to be surveyed in any detail, in the 1970s, by manned submersible, side-scan sonar and other techniques.

The MAR is the 'type example' of the slow-spreading ridge, with a spreading rate varying from 41mm/a in the

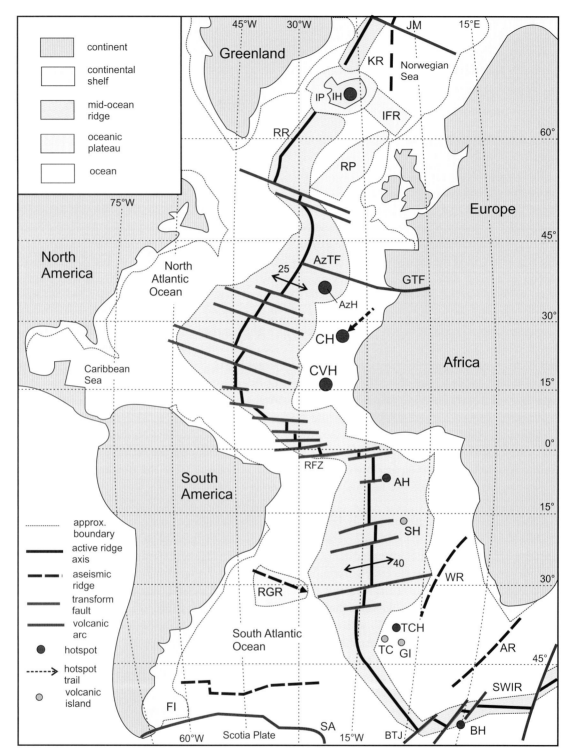

Figure 13.6 The Mid-Atlantic Ridge. Active ridges: KR, Kolbeinsey Ridge; RR, Reykjanes Ridge; MR, Mohns Ridge; SWIR, SW Indian Ridge. Hotspots: AH, Ascension; AzH, Azores; BH, Bouvet; CH, Canaries; CVH, Cape Verde; IH, Iceland; TCH, Tristan da Cunha. Inactive ridges and oceanic plateaux: AR, Aguilhas Ridge; IFR, Iceland–Faeroes Ridge; IP, Iceland Plateau; RFZ, Romanche Fracture Zone; RGR, Rio Grande Rise; RP, Rockall Plateau; WR, Walvis Ridge. Islands: GI, Gough; JM, Jan Mayan; SH, St Helena; TC, Tristan da Cunha. Other features: AzTF, Azores Transform Fault; BTJ, Bouvet Triple Junction; GTF, Gibraltar Transform Fault; SA, Scotia Arc. For the northern end of the Mid-Atlantic Ridge, see Figure 13.2. Based on topographic ocean-floor maps by National Oceanographic and Atmospheric Administration (USA).

southern Atlantic to 18mm/a south of Iceland, and even lower in the Arctic. The topographic profile of a rugged ridge summit with a narrow central rift valley as shown in Figure 13.4B, which was established in these early studies, has been found to be typical of other parts of the MAR as well as of other slow-spreading ridges. The scale of these topographic features is considerable: the ridge as a whole is elevated by about 3000m from the surrounding ocean floor, and the central rift valley, which can be as much as 100km wide, is depressed by over 2000m from the flanking mountains. There is usually a ridge, up to 1km wide and over 100m high, in the centre of the rift, formed by the most recent volcanic eruption. Vertical extensional fractures occur on the floor of the rift valley, whereas normal faulting is characteristic of the flanking mountains (Fig. 13.4).

The half-spreading rate on the western side of the ridge is less than that on the eastern side, at least along the central part of the MAR; that is, the American Plate is moving away from the ridge axis more slowly than either the African or the Eurasian Plates. Looked at another way, relative to the African continent, the ridge axis is moving west at the eastern half-spreading rate while the American continent is moving west faster, at the full spreading rate. The ridge is also moving west relative to both the Azores and Iceland hotspots, which are regarded as being in a (relatively) fixed global frame on timespans of at least 100Ma.

The South Atlantic sector
The southern sector of the MAR (Fig. 13.6) begins at the Bouvet Triple Junction, at 54°S, and keeps to the midline of the South Atlantic Ocean, accommodating to the double bend between North Africa and the Caribbean by means of numerous transform faults that offset the ridge, first in a sinistral, then in a dextral sense, while retaining its trend at right angles to the spreading direction, which varies from close to east–west near the Equator to WNW–ESE in the North Atlantic. The position of the major transform faults is inherited from the original shape of the break-up line when the continents split apart in the Mesozoic. Thus, for example, the Romanche Fracture Zone, which offsets the MAR sinistrally for a distance of 900km, helps to accommodate to the 90° bend in the Brazilian and West African coastlines.

Five islands lie on or near the ridge between Bouvet Island and the Azores: Bouvet Island itself at 54°S, Gough Island and Tristan da Cunha at around 40°S, St Helena at 16°S, and Ascension at 8°S. The Bouvet

Triple Junction lies close to the Bouvet Hotspot. Gough and Tristan da Cunha are part of the same submerged plateau attributed to the Tristan da Cunha Hotspot, which occurs at the intersection of the MAR with the Walvis Ridge, a 3000km-long, rather sinuous feature striking north-eastwards towards the coast of Africa. This feature is thought to result from a deep-mantle plume, which was one of the foci of the line of break-up of Africa and South America in the Cretaceous. As the South Atlantic opened up, two hotspot trails were left, linking the present position of the hotspot to the original sites on the African and South American coasts. A similar ridge on the South American side, the Rio Grande Rise, strikes north-westwards towards Brazil. St Helena is a volcanic island which last erupted 7–10Ma ago and is no longer active. It lay on the ridge at that time but is now offset to the east. Ascension Island is actively volcanic and attributed to the Ascension Hotspot, which is offset about 80km to the east of the ridge axis.

These topographic swellings on or near the ridge are all attributed to mantle hotspots, which have provided enhanced supplies of magma locally into the ridge. The ridge elevation decreases away from these hotspot sites as the magma supply becomes less readily available.

The Cape Verde Islands and the Canaries
The Cape Verde Archipelago, at 17°N, 25°W, and the Canary Islands, at 28°N, 16°W both lie on elevated submarine plateaux and are attributed to hotspots that are located on the African Plate close to the African continental margin. Both groups still contain active volcanoes, but their vulcanicity appears to date back only to the Miocene. Neither has any current connection with the MAR, and both are thought to have remained approximately beneath their present position as the African Plate moved slowly eastwards across them.

The Azores Hotspot
At a latitude of around 39°N, there is a triple junction between the Mid-Atlantic Ridge and the Azores Transform Fault. The nine islands of the Azores lie on a submerged oceanic plateau formed by the Azores Hotspot, which has been volcanically active for the last 7Ma. The volcanic centre currently lies 100–200km east of the present position of the MAR, but it is likely that the ridge lay above the hotspot when it was formed, and subsequently migrated westwards away from it. The ridge itself is appreciably wider at this point. The Azores Transform Fault runs east from the Azores Plateau to

enter the Mediterranean at Gibraltar, where it becomes the Gibraltar Transform Fault (see chapter 4 and Fig. 4.6).

The North Atlantic sector

North of the Azores Triple Junction, between latitudes 40°N and 50°N, there is a relatively short stretch of the Mid-Atlantic Ridge that trends NNE–SSW then, at a point halfway between southern Labrador and Ireland, veers off to the west, and is crossed by a number of WNW–ESE transform faults. The ridge then follows a relatively straight path in a northeasterly direction to Iceland. This part of the MAR is known as the Reykjanes Ridge, and is notable because it is one of the few places on the network where the ridge is oblique to the spreading direction (which is still WNW–ESE).

The island of Iceland lies directly on the ridge and is the only place on the entire MAR where the geology of the ridge can be conveniently examined above ground. Because of its importance, it is discussed separately below.

Iceland to Svalbard

The section of the MAR lying between Iceland and the Svalbard Archipelago is divided into two; the southern part, known as the Kolbeinsey Ridge, continues the northeasterly trend of the Reykjanes Ridge, following a relatively straight path between Greenland and Norway until it reaches a point just north of latitude 70°N, near Jan Mayen Island, where it is displaced eastwards along a major transform fault. Jan Mayen itself lies on a fragment of continental crust left behind when Greenland and Norway separated during the Eocene. It hosts a currently active volcano, Beerenberg, which lies directly above the Mid-Atlantic spreading axis and last erupted in 1970 (Fig. 13.7).

The ridge is displaced eastwards on the transform fault for about 200km, whereupon it continues northeastwards as the Mohns Ridge, until it reaches a point south of Svalbard, where it is again displaced, this time to the west, for over 1000km, and continues across the Arctic Ocean, where it is known as the Gakkel Ridge (see Fig. 13.2).

Figure 13.7 Beerenberg, 2277m, on Jan Mayen Island: the world's most northerly active volcano, last erupted in 1970. Shutterstock©P. Fabian.

The Gakkel Ridge

This submerged ridge follows what is known topographically as the 'Nansen Cordillera' across the Arctic Ocean to the Laptev Sea in Siberia. Despite being obscured by thick permanent sea ice, evidence of ongoing volcanic activity along the ridge was obtained in 1999 from a nuclear submarine and has since been confirmed by scientists operating from ice-breaker vessels. The spreading rate along this ridge is very low, less than 10mm/a; it is classed as an 'ultra-slow' ridge, and it joins a convergent plate boundary that crosses eastern Siberia to form the eastern margin of the Eurasian Plate. The Siberian end of the ridge is termed the Laptev Sea Rift.

Iceland

The island of Iceland offers the ideal opportunity of examining the structures of a typical spreading ridge above sea level. The spreading axis (Fig. 13.8) comes on-shore from the Reykjanes Ridge and runs along the Reykjanes Peninsula, where it is represented by a narrow median rift trough, crossed by a pedestrian bridge with a sign indicating that the American Plate lies on one side and the Eurasian Plate on the other! The ground here is crossed by numerous deep fissures parallel to the spreading axis (Fig. 13.9A). Much of the ground surface here consists of lavas erupted so recently that erosion or vegetation has had little chance to disturb the original volcanic features; examples of ropy and blocky lava illustrate the typical morphology of MOR-type basalts (Fig. 13.9B, C).

Active volcanism is widespread on the island, and provides both geothermal power and district heating for most of the population. The word 'geysir' comes from the famous hot spring of that name near Reykjavik (Fig. 13.10), and volcanic eruptions are common, one of the most spectacular being the Eyjafjallajökull eruption in 2010, whose ash cloud grounded much of Europe's air transport for days.

The island of Iceland has long been thought to lie on a major hotspot, which has resulted in a concentration

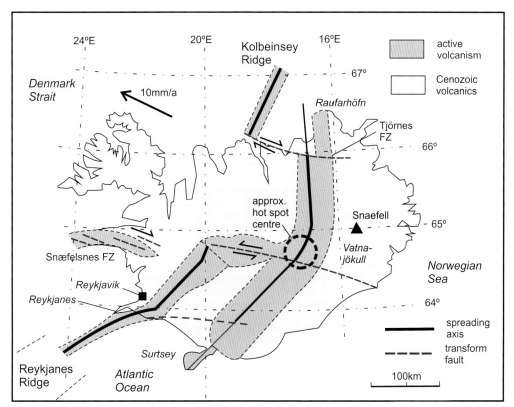

Figure 13.8 Iceland. Distribution of present-day volcanic activity: the whole island consists of volcanic material formed over the last 28Ma. The spreading axis approaching the island from the south, on the Reykjanes Ridge, is offset along a transform fault towards the hot spot, with its centre beneath the Vatnajökull glacier, west of Snaefell Mountain. In the north of the Island, the spreading axis is offset again along a second transform fault to continue north along the Kolbeinsey Ridge. The current volcanic activity on the island is concentrated above the hotspot and along the Reykjanes and Snaefelsnes Peninsulas. After Saemundsson, 1974; and Foulger & Anderson, 2005.

A

B

Figure 13.9 Morphology of Iceland volcanics.
A. Open fissure in the axial rift zone. **B**. Blocky
lava in the foreground with a cracked lava
dome behind. **C**. Ropy lava.

C

Figure 13.10 Geyser. Steam plume from the
hydrothermal vent at the type locality of
Geysir, 40km NE of Reykjavik, SW Iceland.

of volcanism that has enabled this section of the Mid-Atlantic Ridge to be elevated above sea level. Vulcanicity on the island has been continuous since at least 25Ma ago during the late Oligocene, and the currently active rift zone runs through the centre of the island, following a rather complex pattern (Fig. 13.8). This active rift, coinciding with a zone of high heat flow containing many active hydrothermal vents, extends from the Reykjanes Peninsula north-eastwards for 200km towards the centre of the island, where it is offset by 100km along a WNW–ESE transform fault zone towards the southeast. It then continues north-eastwards for a short distance to a point southwest of the mountain of Snaefell, where it turns northwards to continue to the north coast at the Raufarhöfn Peninsula. There, another transform fault, the Tjörnes Fracture Zone, offsets the spreading ridge 100km westwards to connect with the Kolbeinsey Ridge, which continues north-eastwards towards Svalbard. The main focus of volcanic activity, i.e. the 'hotspot', appears to lie beneath the Vatnajökull icecap and rifts seem to have propagated from here both south and northwards beyond the transform faults. Earthquakes associated with the rift zones are consistent with NW–SE extension at right angles to the main rift trend.

The axial rift in the north of the island is about 70km wide and contains flood basalts erupted over the last several hundred thousand years. The currently active belt contains N–S-oriented fissure swarms parallel to the rift margins, along with several individual volcanoes and calderas. The rift is situated within a wide zone where the older lavas have been uplifted into a broad ridge whose flanks are tilted towards the rift.

The island of Iceland sits on a large region of thickened crust that extends from Greenland in the west to the Faeroe Islands in the east (see Fig. 13.6). Beneath central Iceland, the crust is up to 40km thick and is thought to be entirely oceanic in nature. The original hypothesis attributing the enhanced vulcanicity of Iceland to a deep-seated mantle plume fails to explain some of its geochemical and structural characteristics, and the hotspot has more recently been attributed to a more shallow-rooted accumulation of hot mantle material, unconnected with the deeper mantle convective system.

The Indian Ocean

Three separate but linked ridges form the Indian Ocean ridge network: the Southwest Indian (SWIR), Southeast Indian (SEIR) and Carlsberg/Central Indian Ridges (Fig. 13.11). The SWIR and SEIR form the main part of the circum-Antarctic system, the other components being the American–Antarctic and Pacific–Antarctic ridges, discussed below. The Carlsberg Ridge is the best known of the Indian Ocean ridges, and was the first to be investigated in any detail.

The Carlsberg–Central Indian Ridge

This ridge was discovered by the Dana Expedition in 1928–1930 and named after the Carlsberg brewery, the expedition sponsor. Its southern end is the junction with the SW Indian Ridge at 20°S near Rodriguez Island, east of Mauritius. From there it extends north to 10°N near the entrance to the Gulf of Aden, where it connects with the Gulf of Aden and Red Sea rifts (see also Fig. 3.9). The ridge forms the boundary between the African Plate to the west and the Indian Plate to the east. It originated with the separation of India from Africa in latest Cretaceous time and propagated northwards until by the Eocene it had reached the Gulf of Aden, where it was influenced by the Afar hotspot, which initiated the Red Sea and African rift system.

The Carlsberg Ridge is a typical example of a slow-spreading ridge with an average spreading rate of 26mm/a. It is divided into two 'first-order' segments: the more northerly is at right angles to the NE–SW spreading direction, whereas the more southerly one (alternatively known as the Central Indian Ridge) is oriented N–S, at 45° to the spreading direction. Here, the spreading axis is segmented into numerous 'second-order' segments, tens of kilometres in length, which are separated by dextral offset zones that displace the axis by a few kilometres laterally. This geometry enables the spreading axis as a whole to maintain an overall N–S trend while individual segments are orthogonal to the spreading direction. Figure 13.12 illustrates typical along-axis and across-axis profiles obtained from the Carlsberg Ridge. Each segment has an axial valley with one or more volcanic ridges that decrease in height towards the bounding offset zones, which are small fault-bounded basins floored by sediments but devoid of volcanic material.

The Carlsberg Ridge is considered to be typical of slow-spreading ridges, in being characterised by short segment lengths and high topographic relief, compared with faster-spreading ridges, which have lower relief and more symmetrical rift flanks. The difference is ascribed to the greater availability of melts in the case of the faster-spreading examples.

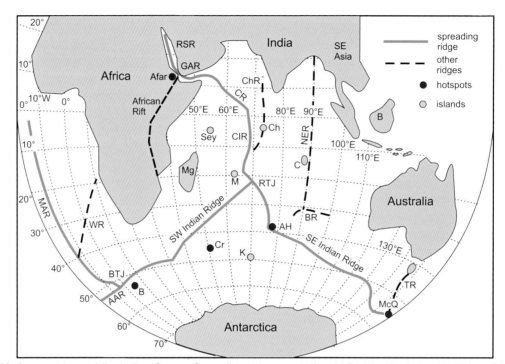

Figure 13.11 Ocean ridges of the Indian Ocean. Spreading ridges and rifts: AAR, Atlantic–Antarctic Ridge; CIR, Central Indian Ridge; CR, Carlsberg Ridge; GAR, Gulf of Aden Rift; MAR, Mid-Atlantic Ridge; RSR, Red Sea Rift. Aseismic ridges: BR, Broken Ridge; ChR, Chagos Ridge; NER, Ninety-East Ridge; TR, Tasman Ridge; WR, Walvis Ridge. Hotspots: AH, Amsterdam–St Paul's; B, Bouvet; Cr, Crozet; McQ, MacQuarie. Triple junctions: BTJ, Bouvet; McQ, Macquarie; RTJ, Rodriguez. Islands: B, Borneo; C, Cocos; Ch, Chagos; K, Kerguelen; M, Mauritius; Mg, Madagascar; Sey, Seychelles. Based on topographic ocean-floor maps by National Oceanographic and Atmospheric Administration (USA).

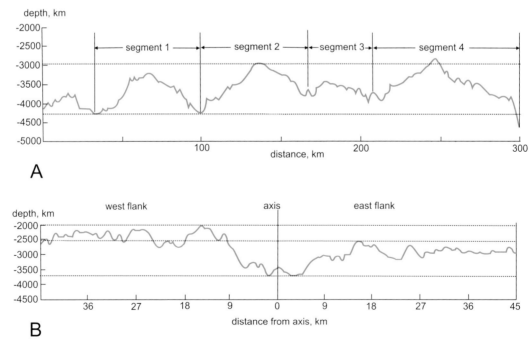

Figure 13.12 Morphology of the Carlsberg Ridge. **A**. Topographic profile along the ridge axis, showing the shape of the ridge segments. Note the *c*.1000m height difference between the ridge crests and the offset basins. **B**. Topographic profile at right angles to the ridge axis showing the difference in height of *c*.500m between the western and eastern flanks. After Murton & Rona, 2015.

The Southwest Indian Ridge (SWIR)

This 7700km-long ridge (Fig. 13.13) is interesting both because of its very slow spreading rate and because it is extending parallel to its length at a faster rate than any of the other ridges. This is due to the fact that the triple points at each end of the ridge are receding from each other, as both the Atlantic and Indian oceans are expanding, and there is no compensating subduction taking place between them. The ridge forms the boundary between the African Plate to the north and the Antarctic Plate to the south, and has a spreading rate varying from 15mm/a (i.e. ultra-slow) to 30mm/a (slow). The ridge was initiated when Africa broke away from Antarctica during the late Cretaceous.

The western end of the ridge is joined to the Mid-Atlantic Ridge and the Atlantic–Antarctic Ridge just west of Bouvet Island, at 54°S, 1°W. The ridge here possesses a deep, wide axial valley, typical morphology of a slow-spreading ridge, and is offset by several NE–SW transform faults. From 10°E to 25°E, the ridge has an overall E–W trend, without transform faults. The western part is characterised by numerous orthogonal segments separated by oblique segments with variable orientations, which are devoid of vulcanicity (Fig. 13.14D). This section has an ultra-slow spreading rate. Between 16° and 25°E, there is a long section of the ridge that is perpendicular to the spreading direction.

From 27°E eastwards, the ridge is offset by numerous transform faults or fracture zones, several of which have very large offsets and are marked by troughs with depths of over 6km. The longest of these is the Andrew Bain Transform Fault with a sinistral displacement of 750km. These transform faults all trend NNE–SSW, parallel to the direction of divergence between the African and Antarctic Plates. Some of the major fracture zones (e.g. the Mozambique Ridge) extend to the continental margin and correspond with sharp breaks in the initial fracturing pattern of Gondwana. Much of this long stretch of the SWIR has quite irregular segmentation. The depth of the axial region here varies from as much as 4730m to 3050m near the Marion Hotspot at 36°E, where there has been an increased magma supply.

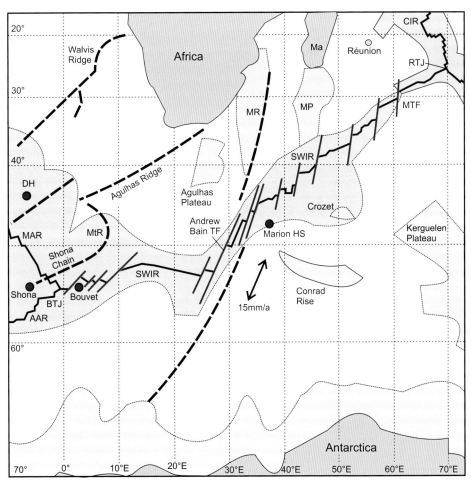

Figure 13.13 The SW Indian Ridge. AAR, American–Antarctic Ridge; BTJ, Bouvet Triple Junction; CIR, Central Indian Ridge; DH, Discovery Hotspot; Ma, Madagascar; MAR, Mid-Atlantic Ridge; MP, Madagascar Plateau; MR, Mozambique Ridge; MtR, Meteor Rise; MTF, Melville Transform Fault; RTJ, Rodriguez Triple Junction; SWIR, SW Indian Ridge. Symbols and colours as in Figure 13.6. Based on topographic ocean-floor maps by National Oceanographic and Atmospheric Administration (USA).

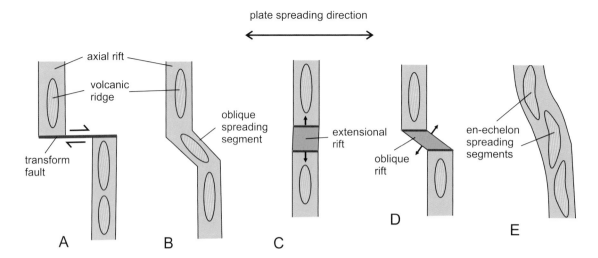

Figure 13.14 Types of discontinuity on spreading ridges. After Searle, 2015.

The easternmost section of the ridge, east of the Melville Transform Fault, is characterised by an ultra-slow spreading rate and is devoid of vulcanicity. The ridge flanks here are rounded and smooth. Divergence is achieved by extensional faulting parallel to the ridge axis rather than by magma injection. The ridge joins the Central Indian Ridge at the Rodriguez Triple Junction, at 70°E. The increase in length of the ridge, required by the increasing separation of the two ends, is achieved by the oblique sections of the ridge, which have experienced a component of NE–SW extension.

The Southeast Indian Ridge (SEIR)

This 6000km-long ridge marks the boundary between the Australian and Antarctic plates, and extends from the Rodriguez Triple Junction in the west to the Macquarie Triple Junction in the east (Fig. 13.15). The ridge was initiated in the Oligocene when Australia and Antarctica began to drift apart. The current average spreading rate is about 65mm/a. The ridge lies close to, and has been influenced by, three hotspots: Amsterdam–St Paul, Kerguelen and Balleny.

The Amsterdam–St Paul Hotspot is a large submerged plateau, situated around Longitude 80°E, 150x200km in extent and elevated 1–3km above the surrounding ocean floor. A wide ridge extends northeast from the plateau, containing a string of submarine volcanoes representing the trail of the hotspot across the Indian Ocean towards Broken Plateau.

The Kerguelen Hotspot is located further southwest, at 70°E. The Kerguelen Archipelago lies on an even larger submerged plateau about 1,250,000km² in extent, situated

within the Antarctic Plate about midway between the SEIR and Antarctica. It has been active volcanically since the early Cretaceous, since when a very large volume of igneous material has been accumulated. The hotspot dates back to the separation of Australia from Antarctica and has been split in two by the subsequent growth of the SEIR. The Australian half of the plateau is known as the Broken Plateau, or Broken Ridge, which extends for 1200km from the southern end of the Ninety-East Ridge towards the southwest corner of Australia. The Ninety-East Ridge extends from Broken Ridge north to the Bay of Bengal, and is considered to mark the track of the hotspot as the Indian Plate travelled north across it towards its present position. The combined Kerguelen–Broken Ridge topographic swell makes it one of the world's largest hotspots, and it is attributed to a class of 'large igneous provinces' (LIPs) attributed to deep-mantle plumes.

About midway between the Kerguelen Hotspot and the Macquarie Triple Junction, around 120°E, the ridge narrows, and is characterised by unusually thin crust, rough topography and anomalously low heat flow. The ridge crest here is more than 2000m deeper than in the adjoining sectors. These properties are thought to indicate a region of mantle downwelling.

The detailed morphology of the SEIR is very similar to the other slow-spreading ridges, such as the Carlsberg and SWIR. A series of first-order segments are separated both by transform faults and by non-transform discontinuities, which are rift-like structures that extend the ridge obliquely to the spreading direction (see Fig. 13.14).

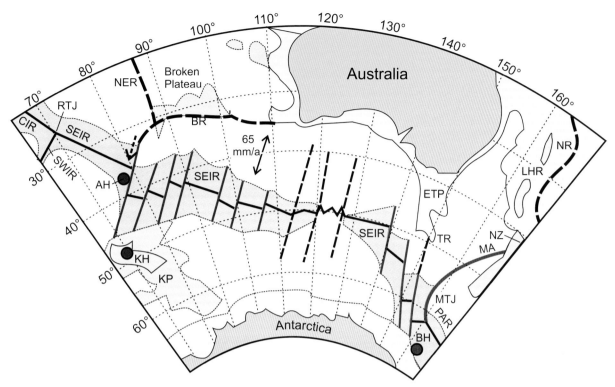

Figure 13.15 The SE Indian Ridge. AH, Amsterdam–St Paul's Hotspot; BH, Balleny Hotspot; BR, Broken Ridge; CIR, Central Indian Ridge; ETP, East Tasman Plateau; KH, Kerguelen Hotspot; KP, Kerguelen Plateau; LHR, Lord Howe Ridge; MA, MacQuarie Arc; MTJ, MacQuarie Triple Junction; NER, Ninety-East Ridge; NR, Norfolk Ridge; NZ, New Zealand; PAR, Pacific–Antarctic Ridge; RTJ, Rodriguez Triple Junction; SEIR, SE Indian Ridge; SWIR, SW Indian Ridge; TR, Tasman Ridge. Symbols and colours as in Figure 13.6. Based on topographic ocean-floor maps by National Oceanographic and Atmospheric Administration (USA).

The easternmost stretch of the ridge is cut by several large transform faults with dextral offsets, which have the effect of changing the overall trend of the ridge from E–W to NW–SE as it approaches the Macquarie Triple Junction.

The Macquarie Triple Junction is the meeting point of the SEIR with the Pacific–Antarctic Ridge to the east and the Macquarie Arc, which forms the western boundary of the Pacific Plate. The triple junction is close to the Balleny Archipelago, which consists of a group of volcanic islands located around 163°E, 67°S. The islands lie on a NW–SE ridge that extends from the SEIR towards the East Tasman Plateau, southeast of Tasmania. They are attributed to the Balleny Hotspot, which is believed to date back to the separation of Australia from Antarctica.

The Pacific
The East Pacific Rise (EPR)

This ridge is the Pacific counterpart of the Mid-Atlantic Ridge, formerly running through the centre of the Pacific

Ocean. However, the reconfiguration of the plates around the Pacific that occurred through the Mesozoic and Cenozoic Eras has resulted in the EPR being displaced eastwards relative to the surrounding continents, to the extent that the Pacific Plate, on the western side of the EPR, now occupies the greater part of the Pacific Ocean (Fig. 13.16) and the EPR has been partially over-ridden by the westward movement of the American Plate. The northern end of the EPR joins the Gulf of California Rift Zone, which separates the Baja California Peninsula from the North American Plate (see chapter 11, Fig. 11.7). This Rift links in turn with the southern end of the San Andreas Transform Fault, which forms the eastern boundary of the Pacific Plate further north.

The northern sector of the EPR runs from the Gulf of California southwards to near the Equator, where it is joined by the Cocos–Nazca Ridge. This sector forms the western boundary of the Cocos Plate, which is being subducted north-eastwards along the Mid-America Trench (see Fig. 12.1). From here, the EPR continues southwards to 35°S, where it meets the transform fault

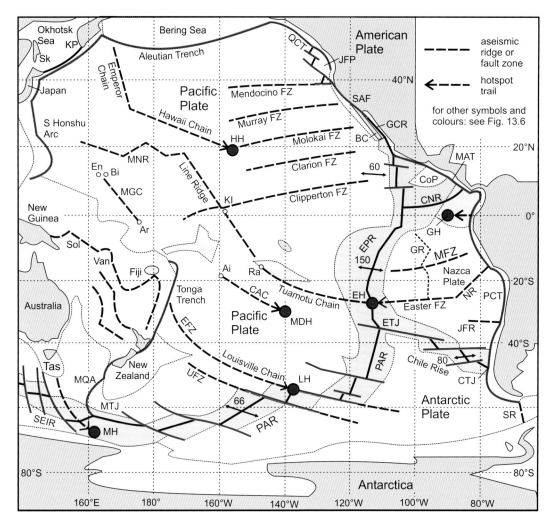

Figure 13.16 The Pacific Ocean ridges. Ridges/rifts (R), transform faults (T) and fracture zones (FZ): CNR, Cocos-Nazca; EFZ, Eltanin; EPR, East Pacific (Rise); GCR, Gulf of California (Rift); GR, Galapagos (Rift); JFR, Juan Fernandez; MFZ, Mendano; MGC, Marshall-Gilbert Chain; NR, Nazca; MNR, Marcus–Necker; PAR, Pacific–Antarctic; QCT, Queen Charlotte; SAF, San Andreas; SEIR, SE Indian; SR, Shackleton; UFZ, Udintsev. Hotspots: EH, Easter; GH, Galapagos; HH, Hawaii; LH, Louisville; MDH, Macdonald; MH, Macquarie. Other tectonic features: CAC, Cook–Austral Chain; CoP, Cocos Plate; CTJ, Chile Triple Junction; ETJ, Easter Triple Junction; JFP, Juan de Fuca Plate; MAT, Middle America Trench; MQA, Macquarie Arc; MTJ, Macquarie Triple Junction; PCT, Peru–Chile Trench. Topographic features: Ai, Aitutaki Atoll; Ar, Arorae Atoll; BC, Baja California; Bi, Bikini Atoll; En, Eniwetok; Ki, Kiritimati (Christmas Island); KP, Kamchatka Peninsula; Ma, Marquesas; Ra, Rangiroa; Sk, Sakhalin; Sol, Solomon Islands; Tas, Tasman Plateau; Van, Vanuatu. Based on topographic ocean-floor maps by National Oceanographic and Atmospheric Administration (USA).

that connects with the Chile Rise, at the Easter Island Triple Junction. This sector forms the western boundary of the Nazca Plate, which is being subducted beneath the Peru–Chile Trench (see Fig. 12.7B).

Most of the EPR is very broad, up to 4000km across, much smoother in profile than the Mid-Atlantic Ridge, and lacks a central rift valley (compare Fig. 13.4A with 13.4B). This difference is attributed to its much greater spreading rate; this reaches 150mm/a in the fastest-spreading sector, near Easter Island, but even

further north, where it is around 60mm/a, it is appreciably faster than the MAR.

The detailed topography of the ridge has been explored in several places and is well illustrated in Figure 13.17, which represents a section of the ridge between latitudes 12° and 13°N. The summit is characterised by a shallow ridge where the current vulcanicity is concentrated. This axial ridge varies from about 8km wide and 250m high at 13°N to 15km wide and between 300m and 400m high at 20°S, on the superfast section

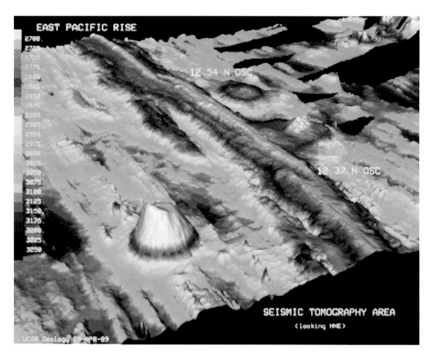

Figure 13.17 Topography of the East Pacific Rise. Map of the seabed on part of the EPR, looking NNE along the ridge between latitudes 12° and 13°N. Depths range from white 2700m to deep blue 3250m. Note en-echelon spreading segments. Science Photo Library, © Dr Ken Macdonald.

of the ridge. The broad sectors, which have a roughly rectangular profile, are attributed to the presence of a shallow inflated magma chamber, whereas the narrower ridges, with a more triangular profile, appear to overlie regions where the magma chamber is deeper, or deflated.

The axial ridge along most of its length possesses a small narrow valley or trough less than 2km wide and about 100m deep. This is thought to indicate the presence of an empty fissure over which the ridge summit has collapsed. In some places the spreading segments are arranged en-echelon (Fig. 13.10E), as shown in Figure 13.17. Not all the vulcanicity is concentrated along the ridge. Occasional circular seamounts have been encountered some distance away from the ridge axis, and several hydrothermal vents have been explored. The morphology of the lava flows is very similar to those of the MAR.

The Galapagos Hotspot

The Galapagos Archipelago is a group of 18 main volcanic islands scattered across the Equator around longitude 90°W, situated on a submarine ridge that extends in an easterly direction towards the coast of Ecuador. The more westerly volcanoes are currently active and the vulcanicity becomes older eastwards, dating back to around 4Ma. The island chain is situated on the Nazca Plate, and is attributed to a hotspot over which the Nazca Plate has tracked, probably since the Miocene. The Galapagos Islands are situated east of the EPR but are linked to it via the Cocos–Nazca Ridge, which forms the boundary between the Cocos and Nazca Plates.

The Easter Island Hotspot

Easter Island consists of three extinct volcanoes forming the summit of a large volcanic mountain rising over 2000m from the ocean floor, and lies at the end of a submarine ridge containing numerous volcanic seamounts. This seamount chain extends for 2700km eastwards, then north-eastwards towards the Chilean coast, where it is currently subducting beneath the Chile Trench. The chain is attributed to the track of, first the Farallon Plate, and more recently the Nazca Plate, as they moved across the Hotspot.

The Pacific–Antarctic Ridge

The southern sector of the EPR, also known as the Pacific–Antarctic Ridge (PAR), runs from Easter Island, first southwards, then gradually bends south-westwards to where it connects with the SE Indian Ridge at the Macquarie Triple Junction (159°W, 55°S), south of New Zealand. This section of ridge forms the boundary between the Pacific Plate to the west and north and the Antarctic plate to the east and south.

At about 140°W, the PAR meets the Louisville Ridge (also known as the Eltanin Fracture Zone), which is a volcanic seamount chain extending for 4300km north-westwards to the Tonga–Kermadec Arc where it is being subducted beneath the Tonga Trench (see chapter 10). This ridge is regarded as the track of the Louisville Hotspot, which is believed now to lie close to the PAR but has left a trail of submerged volcanic seamounts as the Pacific Plate moved across it. Formerly much more vigorous, volcanic activity has decreased since the Oligocene.

Fracture Zones

The Pacific Ocean basin is traversed by a number of aseismic fracture zones. These are expressed topographically in the form of either ridges or rifts, or a combination of these features, and mark the course of formerly active transform faults, which in many cases date back to the Cretaceous Period. The more important of these, shown in Figure 13.16, from north to south, are: the Mendocino, Murray, Molokai, Clarion and Clipperton Fracture Zones, all of which extend westwards from the eastern margin of the Pacific Plate; the Mendanao and Easter Fracture Zones within the Nazca Plate; and the Eltanin–Louisville and Udintsev Fracture Zones, which extend north-westwards from transform faults on the Pacific–Antarctic Ridge. These structures are parallel to the relative movement direction of the plates that they cut. In the case of the North Pacific zones, this direction, which is roughly east–west, tracks the former movement of the Pacific Plate relative to the now-extinct Farallon Plate; the zones in the southern Pacific, which are markedly curved, indicate the current movement path of the Pacific Plate relative to the Antarctic Plate.

The Pacific seamount chains

The vast expanse of the central Pacific Ocean is also crossed by a number of long, relatively straight ridges, some of which have no obvious genetic connection either to the East Pacific Rise or to the volcanic arcs of the western Pacific rim (Fig. 13.16). From north to south, these include the Haiwaii–Emperor Chain, the Marcus–Necker Rise, the Line Islands, the Marshall–Gilbert Island Chain, the Tuamotu Island Chain, and the Cook–Austral Seamount Chain. All of these structures are submerged linear ridges with considerable topographic relief, dotted with small islands and coral atolls. Some of them are clearly hotspot trails, while others have a more enigmatic origin. The best known and most studied example is the Hawaii–Emperor Chain, which has become the type-example of a hotspot chain.

The Hawaii–Emperor Chain

This mostly submerged ridge extends for over 5800km from the Aleutian Trench in the north to the youngest active volcano, the Lo'ihi Seamount, 35km south of the main island of Hawaii. The ridge changes direction abruptly through 120° between the Emperor Seamount Chain and the Hawaiian Ridge. The Emperor seamounts range in age from 85Ma at the northern end to 39Ma at the southern, while the islands on the Hawaiian Ridge are between 28Ma and 7Ma old. The Hawaiian Archipelago itself is still volcanically active; there are five volcanoes on the main island, three of which are still active. The magmas on the main island are tholeiitic rather than alkali-basaltic, and atypical of non-MOR oceanic volcanoes.

The northward increase in age of the volcanoes, coupled with the increasing depth of the eroded seamount tops, led J. Tuzo Wilson in 1963 to propose that the chain represented the trail of a stationary hotspot as the Pacific Plate moved across it – the older, more northerly parts of the ridge becoming gradually deeper as the ridge cooled (Fig. 13.18). The abrupt change in trend was attributed to a change in the direction of travel of the Pacific Plate. Although the model of a stationary plume has been questioned more recently for the Hawaii–Emperor chain, it has acted as a template for explaining many of the other ocean ridges.

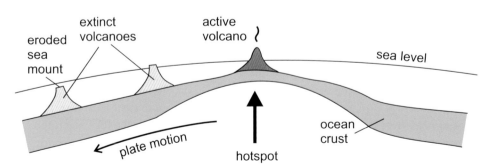

Figure 13.18 Formation of a hotspot trail. Schematic diagram to illustrate the creation of a hotspot trail by a plate moving across a fixed hotspot. Not to scale. After Wilson, 1963.

The Marcus–Necker Rise

This structure is a 4500km-long ridge that extends from around Longitude 150°E in the west, to just south of the Hawaiian Ridge in the east, following an irregular path. The ridge is broad, up to 1000km in places, and elevated by 300–500m from the ocean floor. It contains many submerged peaks and several small islands, but differs from the other seamount chains in not following a NW–SE linear track, and its origin is unclear.

The Line Islands

These are a chain of atolls situated on volcanic seamounts that form a 4800km-long NW–SE ridge extending from around 20°N to well south of the Equator. They include Kiritimati, formerly known as Christmas Island. Volcanic ages range from mid-Cretaceous to late Eocene, becoming younger southwards. Although a hotspot origin has been proposed for the chain (see below), the age distribution does not appear to match that of a simple hotspot trail.

The Marshall–Gilbert Island chains

These two archipelagos form part of a largely submerged ridge that extends from Eniwetak Atoll, at 11°30'N, 162°20E to Arorae Atoll, at 2°38'S, 176°49'E, at the southern end of the Gilbert Chain. The Marshall Islands are scattered over a large area, whereas the Gilbert Islands form a more linear chain. Both groups of islands are composed of coral atolls situated on volcanic seamounts.

Bikini Atoll, near the northern end of the Marshall Islands, achieved notoriety because of the atomic bomb testing carried out there by the USA between 1946 and 1958, which involved the removal and resettlement of the population. Recorded volcanic ages range from late Jurassic in the north to early Cretaceous in the south. The Marshall–Gilbert Ridge is approximately aligned with the Cook–Austral Chain further southeast and may represent the proximal part of the same hotspot track.

The Tuamotu Ridge

The Tuamotu Archipelago consists of nearly 80 islands and atolls situated on a broad submerged ridge aligned NW–SE. The large island of Rangiroa, at 15°07'S, 147°38'W is near the northern end of the chain. The ridge is aseismic and there are no recorded volcanic eruptions. The ridge is aligned with the Line Islands further northwest and both groups have been attributed to the Easter Hotspot, which lies on the East Pacific Rise further southeast.

The Cook–Austral Seamount Chain

This chain, consisting of 13 islands and atolls, extends for 2200km south-eastwards from the island of Aitutaki in the Cook Islands to the Macdonald Seamount, which is a currently active volcano located at 140°W, 30°S. Like the other NW–SE seamount chains, it is aligned parallel to the direction of Pacific Plate motion and is attributed to the Macdonald Hotspot.

14

Older Mountain Belts

The previous ten chapters have described the system of mountain belts and associated tectonic features that have developed over the last 200Ma during the Mesozoic and Cenozoic Eras and have resulted in the present-day distribution pattern of active seismicity and vulcanicity. The processes that gave rise to this pattern are in most cases relatively easy to reconstruct by tracking successive plate movements using the magnetic stripe data from the ocean floors, as explained in chapter 3. However, the oldest ocean crust that can be dated in this way is of Jurassic age, which means that the starting point for the reconstruction of Mesozoic–Cenozoic plate movements is the Pangaea Supercontinent at the end of the Palaeozoic, as shown in Figure 14.1. However, the geological evidence for the previous existence of mountain belts throughout most of Earth history is clear, and this chapter gives examples of some of the pre-Mesozoic belts that are recognised, and of the different methods used to reconstruct them in the absence of ocean-floor data.

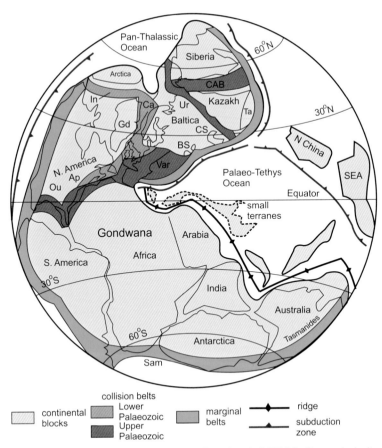

Figure 14.1 Reconstruction of the continents during the late Permian (c.250Ma). Orogenic belts: Ap, Appalachian; Ca, Caledonian; CAB, Central Asian; In, Inuitian; Ou, Ouachita; Sam, Samfrau; Ur, Urals; Var, Variscan. Continental blocks: Ka, Kara; Kazakh, Kazakhstania; SEA, SE Asia; Ta, Tarim. Geographic features: BS, Black Sea; CS, Caspian Sea; Gd, Greenland. After Cocks & Torsvik, 2006.

Methods of reconstruction of older mountain belts

Old mountain belts may or may not betray their presence in the form of topographic elevations, and the older they are the more chance that the effects of erosion have reduced or even completely removed any topographic expression. The investigation of pre-Mesozoic belts has therefore focused on the evidence for the various geological processes that are involved in the creation of a mountain belt – these orogenic processes that collectively are responsible for the creation of orogenic belts.

A major advantage of studying older orogenic belts is that erosion has made it easier to access the deeper parts of such belts and therefore to gain information about tectonic processes in the deep crust. It should not be surprising that much of the information about ductile deformation processes and the metamorphic conditions that accompany them have been gained from regions such as the Scottish Highlands.

The main over-riding processes involved in the formation of an orogenic belt are subduction and collision. Evidence of subduction may be in the form of obducted ophiolite complexes, or of the magmatic products of subduction-related volcanism. Evidence of collision may be in the form of differing geological histories on either side of a suture line, or of the structural and metamorphic effects of crustal thickening and deformation. It is important to recognise, however, that reconstructions of former orogenic belts invariably involve a degree of speculation, which becomes increasingly more significant the older the belt. Intense disagreements have accompanied the study of many of these older belts, and interpretations are being continually revised. Accordingly, all reconstructions should be viewed with a degree of scepticism!

The Palaeozoic world

At the end of the Palaeozoic Era, during the late Permian Period, almost all the continental masses were arranged together as shown in Figure 14.1, in the supercontinent of Pangaea. The geological evidence for this arrangement was briefly discussed in chapter 3. Had geological observers been around then, 250Ma ago, several great mountain ranges, rivalling the Himalayas in scale, would have been evident. These are of two types: collisional belts traversing the supercontinent where the various component parts had come together; and subduction-related belts around the margins. There are three main collisional belts: the Caledonian–Appalachian–Variscan belt along the join between North America on one side, and Europe

and Africa on the other; the Urals between Europe and Asia; and the Central Asian Belt between Siberia and Khazakhstania.

The circum-Pangaea belt follows the present western boundary of the Americas, and is the precursor to the Cordilleran and Andean chains of the present day. The Palaeozoic history of these belts has been largely obscured by the Mesozoic–Cenozoic orogenic activity responsible for the existing mountain belts. Another section follows the outer margin of Gondwana, with representatives in southernmost Africa, western Antarctica and eastern Australia, where it is known as the Tasman Belt. The existence of this orogenic belt was recognised in the early twentieth century as evidence for continental drift by Wegener and Du Toit and referred to as the 'Samfrau Geosyncline'. The latter belt is relatively unaffected by subsequent orogenesis.

The collisional belts that cut through the Pangaea Supercontinent record how the various continental masses that preceded it were assembled to form the supercontinent. The main continental units that existed during the early Palaeozoic are shown in Figure 14.2; the largest of these by far was Gondwana, which comprised most of the present continents of South America, Africa, Australia, Antarctica and India. The other continents were Laurentia, which consisted of the core of North America plus Greenland; Baltica, which represents the old Precambrian core of Europe, and includes Scandinavia and much of east-central Europe; and Siberia, the Precambrian core of Asia. There were also numerous smaller continental terranes that had split off from the major continents and travelled independently across the oceans.

The Palaeozoic history of the major continents has been reconstructed with a fair degree of accuracy by means of palaeomagnetic information from within each of the various continental units, supported by geological indicators of climatic conditions. Palaeolatitudes can usually be closely constrained by this means but there is no method of determining palaeolongitude, so that the exact position of a continent on a particular palaeolatitude can only be guessed.

The assembly of Pangaea

During the Cambrian Period, a subduction zone lay along the margin of Gondwana facing west towards Baltica and Siberia, and as a consequence, back-arc rifting caused several continental fragments to be split off from the Gondwanan margin (Fig. 14.2A). By the

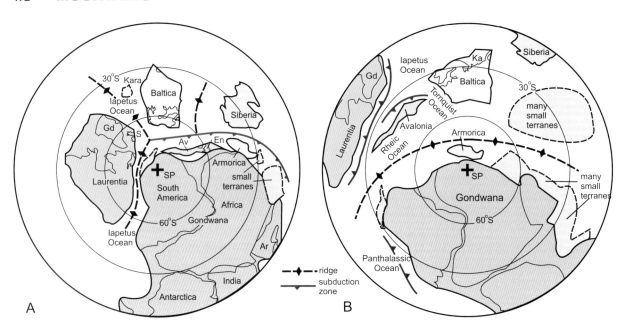

Figure 14.2 Reconstruction of the continents, I: **A**, at the end of the Precambrian (c.550Ma); **B**, in the mid-Ordovician (c.460Ma). Av, Avalonia; En, England; Gd, Greenland; S, Scotland; SP, South Pole. After Cocks & Torsvik, 2006.

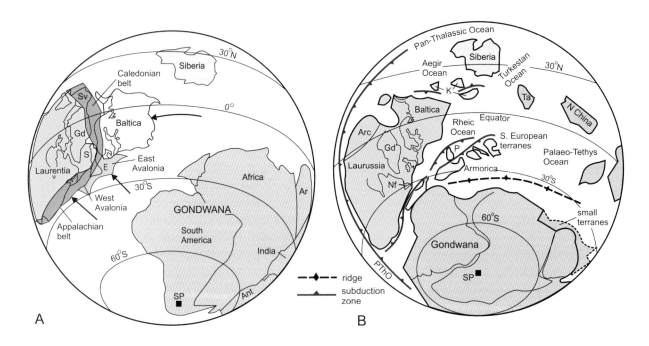

Figure 14.3 Reconstruction of the continents, II: **A**, in the Silurian (c.420Ma) showing the Caledonian–Appalachian belt (in blue) resulting from the collision with Baltica in the northern sector and East and West Avalonia in the southern. Arrows show the convergence direction. Note the positions of Scotland (S) and England (E); **B**, in the late Devonian (c.370Ma). The small terranes in green are destined to collide with Laurussia to form the Hercynian Belt, the blue terranes combine to form Kazakhstania, which collides with the Baltic margin of Laurussia to form the Urals Belt. Arc, Arctica; E, England; Gd, Greenland; K, Kazakhstania; Nf, Newfoundland; P, Perunica; S, Scotland; Sv, Svalbard. A, after Dalziell, 1997; B, after Cocks & Torsvik, 2006.

mid-Ordovician, one of these fragments, Avalonia, had become detached from Gondwana and had travelled independently towards Laurentia and Baltica (Fig. 14.2B). The subsequent collision of Avalonia with Laurentia during the Silurian gave rise to the southern sector of the Caledonian–Appalachian Belt (Fig. 14.3A). During the same period, Baltica was also moving towards Laurentia and collided with it during late Silurian to early Devonian times to create the Scandinavian sector of the Caledonides. As a result of this collision, the western margin of Baltica came into contact with Avalonia to form the eastern branch of the Caledonides (now obscured by the later Hercynian Belt), which traverses Europe from Denmark to the Black Sea (see Fig. 14.1). On the northern side of Laurentia, the small continental terrane of Arctica collided with Laurentia to form the Inuitian Belt, which met the northern end of the Caledonian Belt at Svalbard.

During the Devonian, several other terranes became detached from Gondwana and travelled northwards towards the combined continent of Laurentia–Baltica, now known as 'Laurussia' (Fig. 14.3B). During the Carboniferous, these terranes successively amalgamated with the Baltic and Laurentian margins to create the Hercynian Orogenic Belt (shown in brown in Fig. 14.1), which extends across central and southern Europe (where it is known as the Variscan) and along the southern Appalachians. The last two episodes in this assembly process were the collision of the West African margin of Gondwana with southeastern Laurentia to complete the Southern Appalachian Belt, which occurred in the late Carboniferous, and the collision of Khazakhstania and then Siberia with eastern Baltica to create the Urals Belt, which was finally completed in the early Permian to create the Pangaea assemblage as shown in Figure 14.1.

Of the Lower Palaeozoic orogenic belts, the Caledonides is the one that has been most intensively studied and on which much of the research into orogenic processes has been concentrated. Two contrasting sectors are discussed: Scandinavia–Greenland and British Isles–Newfoundland.

The Scandinavian Caledonides

The Caledonian Orogenic Belt extends over 1700km along the Atlantic coast of Scandinavia (Fig. 14.4) and a matching belt of a similar length lies along the east coast of Greenland. Figures 14.2B and 14.3A show how this orogenic belt resulted from the convergence of Baltica with Greenland during the Ordovician and their ultimate collision in the Silurian. The belt was subsequently split in two by the opening of the North Atlantic Ocean in the early Cenozoic.

The Scandinavian mountain chain occupies a belt varying in width from about 250km in the north to nearly 500km in the south. There are many peaks over 2000m in height, the highest being Galdhøpiggen, 2469m, on the Jotunheimen Plateau. The orogenic belt is narrower, varying from about 100km to 400km in width, and occupies a coastal belt 1700km long, ending on the north coast of Norway, and reappearing in the Svalbard archipelago in the Arctic Ocean, a further 500km to the north.

Structure and composition of the Scandinavian belt

In mainland Scandinavia, the belt consists of a series of nappe complexes divided into four tectonic zones, each of which has been thrust eastwards onto the Baltica craton (Fig. 14.4). The lowest of these zones, known as the Lower Allochthon, consists of nappes containing a mid-Proterozoic basement, similar to that of the foreland, overlain by a late Proterozoic to Lower Palaeozoic shallow-marine platform cover. This sedimentary cover is exposed mainly in the east of the orogenic belt, whereas the outcrops of the Precambrian basement appear in a series of basement 'windows', the largest of which is a coastal area about 100km wide extending from Trondheim in the north to Bergen in the south.

The overlying Middle Allochthon zone consists of greywacke sandstones intruded by a dolerite dyke swarm, and interpreted as the thinned passive continental margin of Baltica with its overlying continental slope sediments. It is overlain by the Upper Allochthon zone consisting of a late Proterozoic to Lower Palaeozoic clastic sedimentary sequence containing a calc-alkaline volcanic suite, and includes an ophiolitic assemblage. This unit is interpreted as the product of an intra-oceanic volcanic arc produced by a subduction zone dipping towards Laurentia rather than Baltica, since there is no sign of a volcanic arc on the Scandinavian side. However, the Ordovician to Silurian sediments contain Scandinavian rather than Laurentian faunas, indicating their relative closeness to Baltica.

Differences in the age of the metamorphic event accompanying the arc emplacement suggest that two different arc collisions may have occurred: one in the north at $c.505$Ma, termed the Finnmarkian event, and one in the central sector at $c.493–482$Ma, termed the Trondheim event.

Figure 14.4 Tectonic units of the Scandinavian Caledonides. All boundaries between the various units are thrusts. The cross-section is taken across the central part of the orogen along the line A–B. LA, Lower Allochthon; MA, Middle Allochthon; UA, Upper Allochthon; UMA, Uppermost Allochthon. G, Galdhøppen (Mountain); JP, Jotunheimen Plateau. After Gee et al., 2010.

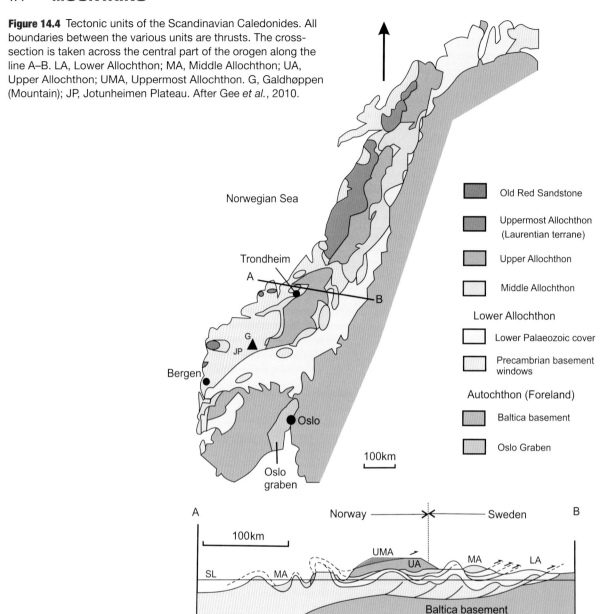

The structurally highest zone, known as the Uppermost Allochthon, is composed of a quite different assemblage, which is exotic to Scandinavia and considered to be derived from Laurentia (i.e. Greenland). This unit is composed of a basement of gneisses and schists overlain by Ordovician to Silurian sediments containing Laurentian faunas, together with volcanics. It is considered to have been formed initially by the overthrusting of a volcanic arc onto the Laurentian margin, then subsequently backthrust onto the Baltic continent as a result of the main Laurentia–Baltica collision (the Scandian event) during the Silurian.

Coarse clastic deposits of late Silurian to early Devonian age occur in intra-montane sedimentary basins at several places along the west coast. These formed as a consequence of syn-orogenic and post-orogenic gravitational collapse, which prompted extensional movements on some of the pre-existing thrusts.

Tectonic history

This is summarised in Figure 14.5. Sedimentation commenced on the western margin of Baltica during the late Proterozoic, on a mid-Proterozoic basement of gneisses and granites, with a sequence of continental and

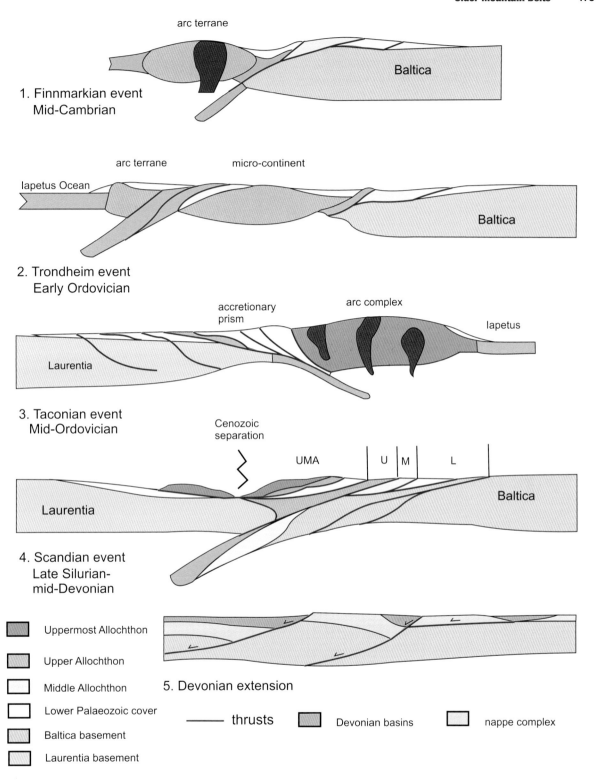

Figure 14.5 Tectonic evolution of the Scandinavian Caledonides: interpretative cartoon profiles. UMA, Uppermost Allochthon; U, Upper Allochthon; M, Middle Allochthon; L, Lower Allochthon. **1**, Finnmarkian event: collision of microcontinent and arc terrane in the northern sector of the Baltica margin. **2**, Trondheim event: collision of volcanic arc terrane in the central sector. **3**, Taconian event: collision of volcanic arc terrane with Laurentia. **4**, Scandian event: collision of Laurentia and Baltica and overthrusting of Laurentian Uppermost Allochthon onto Baltica. **5**, Devonian extensional faulting and creation of sedimentary basins in half-graben. After Roberts, 2003.

shallow-marine sediments, including glacial deposits of the Varangian Glaciation at *c.*668Ma.

For most of the Cambrian Period, the western margin of Baltica was a stable platform covered by shallow-marine deposits, typically of black shales and carbonates. During the Cambrian, at least two volcanic island arcs formed outboard of the Baltic margin, within the Iapetus Ocean, above a west-dipping subduction zone, and accretionary prisms developed within their foredeep basins. In the late Cambrian, one of these arcs collided with the Scandinavian margin during the Finnmarkian event (Fig. 14.5.1).

In the earliest Ordovician, black shale deposition continued on the Baltic shelf; however, a widespread unconformity marked an episode of uplift at about 478Ma, during the Trondheim event, which is ascribed to the collision of a second volcanic arc with the Baltic margin (Fig. 14.5.2). Differences of at least 20Ma in the age of the metamorphism associated with these events suggest that two different arc collisions may have occurred, the earlier one in the north, and the later in the central sector of the orogen.

A very similar event occurred rather later, between 470 and 465Ma, in the mid-Ordovician, although in this case the arc was obducted onto the Laurentian margin and is correlated with the widespread Taconic event in North America (Fig. 14.5.3). In the later Ordovician, gradual mixing of the Baltic faunas with Laurentian species indicates the approach of the two continents.

By the end of the Ordovician, the closure of the Iapetus Ocean separating the two continents was signalled by the emplacement of the Laurentian basement of the Uppermost Allochthon onto the Baltic margin (Fig. 14.5.4). The early Silurian limestones of the foreland are overlain by shales and turbiditic greywackes that are interpreted as the products of a foreland basin created by the depression of the Baltic margin by the Upper Allochthon. By the mid-Silurian, the nappes of the Upper Allochthon had been transported eastwards beneath the Uppermost Allochthon and were subjected to metamorphism, while greywacke deposition continued in the foreland basin. These deposits were succeeded by non-marine clastic sediments as the nappe complex was uplifted and eroded. Translation of the Upper and Middle Allochthons across the foreland basin continued into the late Silurian, incorporating elements of the basin into the complex. The nappes of the Lower Allochthon were formed in advance of the Middle Allochthon, and involved units of the foreland platform cover.

Movements in the nappe complex continued into the early Devonian. During this period, the Atlantic coastal zone was uplifted into a basement antiform due to isostatic rebound of the crust in response to the removal of the overlying nappes. This removal was due partly to erosion, but was mainly a response to an extensional reversal of the thrust-sense shear zones causing gravitational sliding of the nappes away from the tectonically thickened central zone of the orogeny (Fig. 14.5.5). The uplift revealed basement rocks that had experienced high-temperature and high-pressure metamorphism caused by their having been underthrust beneath the Laurentian continental margin during the Baltica–Laurentia collision.

The present-day height of the mountain range, together with a crustal thickness of 40–45km, indicate that the orogenic belt has been reduced in scale by perhaps 50% compared with modern examples such as the Alps. Nevertheless, the Scandinavian Caledonides are still an impressive crustal structure despite the 400Ma or so of uplift and erosion that have affected them since the Devonian.

The Caledonides of the British Isles and Newfoundland

The Caledonian orogenic belt, or Caledonides, extends southwards from Norway and East Greenland, through the British Isles to Newfoundland and eastern North America (Fig. 14.6) where it merges with the Appalachian Belt. The original extent of the belt only becomes obvious when the effects of the opening of the Atlantic Ocean in the Cenozoic Era have been restored. Figure 14.3A showed a reconstruction of the relative positions of the continents during the mid-Palaeozoic, from which the Caledonian–Appalachian orogenic belt (shown in blue) is seen to represent a collisional orogen between Laurentia (North America plus Greenland), Baltica (northern mainland Europe) and two (possibly joined) microplates, East and West Avalonia. According to this reconstruction, the northern part of the British–Irish sector has resulted from collision between Baltica and Laurentia, whereas the southern part, including England and Wales, is ascribed to collision with the Avalonia microplate that has migrated from Gondwana.

Regional context

Figure 14.6 is an interpretation of how the zones of the British–Irish sector of the Caledonides might link up with Norway and Greenland to the north and

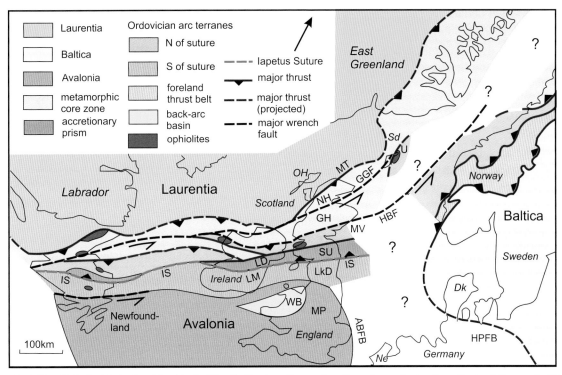

Figure 14.6 The Caledonide belts of the North Atlantic Region. Reconstruction after restoring the effects of Cenozoic Atlantic opening. Note the major NW-directed thrusts north of the suture and SE-directed thrusts south of the suture. ABFB, Anglo-Brabant Fold Belt; Dk, Denmark; GGF, Great Glen Fault; GH, Grampian Highlands; HBF, Highland Boundary Fault; HPFB, Heligoland-Pomerania Fold Belt; IS, Iapetus Suture; LD, Longford Down Massif; LkD, Lake District; LM, Leinster Massif; MP, Midlands Platform; MV, Midland Valley; Ne, Netherlands; NH, Northern Highlands; MT, Moine Thrust; OH, Outer Hebrides; Sd, Shetland; SU, Southern Uplands; WB, Welsh Basin; U, Unst ophiolite. Note orientation of North. After Leslie *et al.*, 2008; and Dewey & Shackleton, 1984.

Newfoundland to the south. The Northern Highland and Grampian Highland zones of Scotland are part of a regional metamorphic core complex, represented also in Ireland, Newfoundland and East Greenland, which has been thrust to the northwest onto the Laurentian continent. In Newfoundland, Ireland and Shetland, this zone is overthrust on its southeastern side by Ordovician arc terranes.

South of these zones are southeast-directed units, which include Ordovician accretionary prisms in Newfoundland, Scotland and Ireland, and volcanic arc terranes on the northern margin of Avalonia. Between Avalonia and Baltica, there are two further branches of the Caledonides: the Anglo-Brabant Fold Belt, which defines the eastern margin of Avalonia, and the Heligoland–Pomerania Fold Belt, which follows the western margin of Baltica. These two belts, which are separated by a narrow stable block, are not well known, being almost completely obscured by younger cover.

An important role is played by major strike-slip faults – the Great Glen, Highland Boundary and Southern Uplands faults being only the more obvious. The total sinistral strike-slip displacement on these is unknown, but estimates have ranged from a few hundred to over 1000 kilometres. Consequently, none of the terranes within the Caledonian Belt can be directly linked to its neighbour, which makes interpretation difficult.

The strike-slip faulting is a result of the late Silurian to early Devonian collision between Laurentia and Avalonia, which must have been oblique to the plate boundary such that the convergence direction was partitioned into components at right angles to, and parallel to, the boundary. The earlier (mid-Silurian) collision between Laurentia and Baltica seems to have been more nearly at right angles to the Baltica plate margin. The effects of this Scandian event on the British Isles south of the Northern Highlands are not obvious. This may be the result of the movements on the Great Glen Fault, which have juxtaposed terranes that were previously far to the southwest, away from the influence of the Scandian collision.

The British–Irish Caledonides

This sector of the Caledonides occupies the key position at the centre of the Caledonide Belt, where the three separate branches meet, and provides the most complete cross-section through the belt. It has also benefited from intensive study by several generations of geologists since the late nineteenth century.

The belt here is about 300km wide and can be divided into nine tectonic zones separated by major fault boundaries (Fig. 14.6). These are, from northwest to southeast: the Northwest Foreland (i.e. Laurentia), the Moine Thrust Belt, the Northern Highlands, the Grampian (or Central) Highlands, the Midland Valley, the Southern Uplands, the Lake District, the Welsh Basin and the Midlands Platform, or Southeast Foreland, part of the East Avalonia microplate. The Northern and Grampian Highlands zones together belong to the central metamorphic core of the orogenic belt. The zones were established in Scotland and England; however, the Northern Highlands, Grampian Highlands, Southern Uplands and Lake District zones can also be traced across into Ireland, though with considerable differences in detail.

The Northwest Foreland

This zone consists of a mainly gneissose basement of early Proterozoic age (the Lewisian Complex) comparable with the formerly adjacent early Proterozoic belt in East Greenland at the southeastern margin of the Laurentian continent. This basement is unconformably overlain by continental red-bed sequences of late Proterozoic age, and by a Cambrian to early Ordovician shallow-marine shelf sequence (Fig. 14.7.1–2).

The Moine Thrust Belt

Here the foreland sequence is involved in several major thrust packages, each of which is divided internally by smaller thrust slices. These are overlain by an allochthonous nappe consisting of the Moine Complex of the Northern Highlands, resting on the Moine Thrust (Fig. 14.8A). Elongation lineations indicate a WNW-directed shear sense. The present outcrop width of the zone is very narrow – from less than a metre to a maximum of only 10km. However, the zone must originally have extended for a considerable distance both eastwards beneath the Moine thrust, and also westwards to incorporate the Caledonian thrusting in the basement of the Outer Hebrides, where the platform cover has been removed by erosion.

The thrusts seem mainly to have propagated forwards, towards the foreland, such that the youngest thrusts, involving only Cambrian sediments, carry older Lewisian basement on their roofs, with the oldest, the Moine Thrust, overlying the whole package. The Moine Thrust differs from the younger thrusts in being characterised by a thick band of mylonite, indicating derivation from considerable depth. The thrust movements are attributed to the mid-Silurian Scandian orogenic phase.

The Moine Thrust Belt is well exposed in the Assynt district of NW Scotland, which lies in the North West Highlands Geopark, and is known internationally as the area where Ben Peach, John Horne and their colleagues of the British Geological Survey (Peach *et al.*, 1907) first mapped and explained the complex thrust geometry in the latter part of the nineteenth century. Much of the geology can be readily appreciated in the scenery from the roadside, and some of the key exposures are explained at viewpoints and at a visitor centre at Knockan Crag, where a section through the Moine Thrust is exposed.

The Northern Highlands

This zone contains the Moine Supergroup, which consists of a thick sequence of late Proterozoic marine clastic sediments. They rest on a basement of Lewisian gneisses, and are overlain in the east by post-orogenic Devonian cover (the Old Red Sandstone). The Moine Supergroup has been intensely deformed and metamorphosed, and has experienced three separate orogenic events. The earliest of these, the Knoydartian, took place around 800Ma and is represented by granitic intrusions, pegmatites and metamorphic ages. However, the nature and extent of the Knoydartian structures have been obscured by the younger orogenic phases. The second phase, the Grampian orogeny, is the main tectonic event to affect the Grampian zone to the south, but its effects in the Northern Highlands are less obvious. The Scandian orogeny, attributed to the Silurian collision between Baltica and Laurentia, is the most important tectono-thermal event to affect the northern part of the Northern Highlands. Scandian structures include ductile thrusts and recumbent folds that are overthrust towards the west-northwest (Fig. 14.7.3). These structures were re-folded by more upright folds with a NE–SW to NNE–SSW trend. Folding was accompanied by high-grade metamorphism dated at 435–420Ma (mid- to late Silurian); these dates also correspond to the date of the movements on the Moine Thrust.

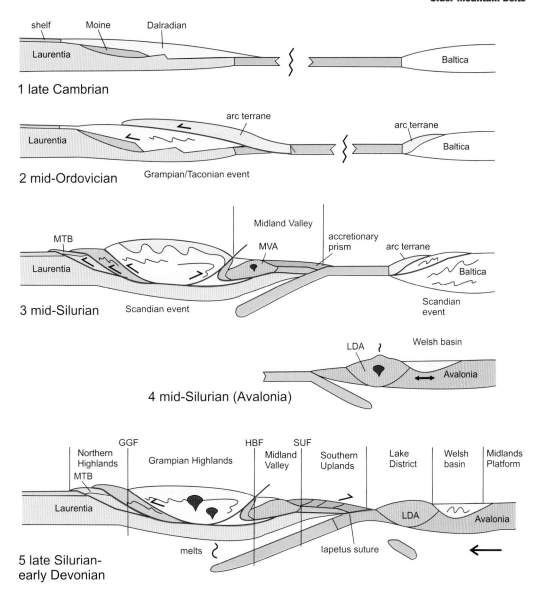

Figure 14.7 Tectonic evolution of the British–Irish Caledonides: cartoon profiles. Note that these are composite profiles as no direct correlation is possible because of the large displacements on the major strike-slip faults. GGF, Great Glen Fault; HBF, Highland Boundary Fault; LDA, Lake District Arc; MTB, Moine Thrust Belt; MVA, Midland Valley Arc; SUF, Southern Uplands Fault.

The Grampian (Central) Highlands

This zone contains a thick marine clastic sequence, the Dalradian Supergroup, thought to have been laid down on the extended passive margin of Laurentia in several half-graben (Fig. 14.7.1). The base of the sequence is of late Proterozoic age; towards the top, greywacke deposits and mafic lavas culminate in a Lower Cambrian limestone. The upper part of the sequence is therefore the deeper-water equivalent of the shelf deposits of the

foreland. The basement in the west is composed of early Proterozoic gneisses, but in the northeast, the Dalradian rocks lie on gneisses that are correlated with the Moine Supergroup.

The earliest deformation to affect the Dalradian rocks consists of NW-directed ductile thrusts and recumbent folds; these are refolded by overfolds that are also NW-directed in the northwest part of the zone, but in

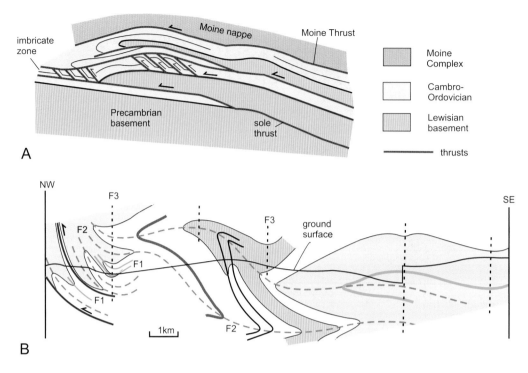

Figure 14.8 Caledonian structures: (**A**) in the Moine Thrust Belt, and (**B**) in the Grampian Highlands. **A**. Note the imbricate structure of the lower middle nappe, consisting of thin slices of Cambrian strata sliding on a weak layer of shales. The higher nappes were emplaced first and deformed as the lower nappes progressed forwards. **B**. This cross-section through the central part of the Grampian Highlands shows the earliest recumbent folds (F1) refolded by younger overfolds (F2) and finally by later upright folds (F3). A, after Elliott & Johnson, 1980; B, after Treagus, 2000.

the southeast are flat-lying and appear to be overthrust towards the southeast (Fig. 14.8B). Later folds are more upright with a NE–SW trend. The main regional metamorphism, dated at *c.*475–460Ma, accompanied the earlier folding and is attributed to the mid-Ordovician Grampian event. The later folds were formed during retrogressive metamorphism.

The Grampian event is attributed to a collision between Laurentia and an oceanic arc terrane represented, in part, by the Midland Valley Zone. The presence of a thick ophiolite sequence in eastern Shetland, which includes a substantial upper mantle component, has prompted the suggestion that the Dalradian was overlain by a large ophiolite nappe, now removed by erosion, emplaced prior to the collision (Fig. 14.7.2). The presence of extensive ophiolite sequences in Newfoundland and Norway is evidence that oceanic arc terranes of this type were regionally important (see Fig. 14.6). The later folds in the southeastern part of the zone are attributed to the collision with Avalonia in the Silurian.

A series of syn-orogenic granite plutons intruded in the period 470–460Ma are attributed to crustal melting in the thickened orogen during the Grampian event.

However, the many large post-orogenic granite plutons, such as the famous Glencoe and Ben Nevis complexes, are linked with a later (Silurian) episode of subduction (Fig. 14.7.5), discussed below.

The Midland Valley

Lower Palaeozoic rocks are only exposed in a few small inliers near the northern and southern margins of the Midland Valley, the Caledonian history of which is mostly concealed by Upper Palaeozoic cover. The oldest rocks in the inliers are of lowermost Ordovician age and consist of an ophiolite assemblage succeeded by Ordovician to mid-Silurian sediments containing volcanic clasts ranging up to boulder size. The sediments are unmetamorphosed and gently folded. The Midland Valley is interpreted as a Lower Palaeozoic oceanic volcanic arc terrane (Fig. 14.7.3), better represented in Norway. Fossil assemblages both here and in the Southern Uplands to the south have Laurentian rather than Gondwanan affinities.

Geophysical evidence indicates that the Ordovician cover lies on a crystalline basement similar in properties to that underlying the Dalradian to the north, and in

Ireland, ophiolites are thrust over gneissose basement, which is likely therefore to belong to the stretched margin of Laurentia. These ophiolites, together with the more extensive oceanic assemblages in Newfoundland, East Greenland and Norway that have been thrust over the Laurentian and Baltic forelands respectively, are thought to represent former back-arc basins.

Major sinistral faults

Both the Great Glen and Highland Boundary Faults have experienced large lateral displacements, since the rocks on each side do not match up. Estimates of the amount of displacement on the Great Glen Fault vary from *c.*160km to over 500km, but the presence of Moine rocks with similar dates on both sides of the fault suggests that it is probably not a major terrane boundary. However, the sequences presently juxtaposed across the Highland Boundary Fault cannot be directly matched, indicating that this fault is a terrane boundary representing a suture zone between Laurentia and a separate Midland Valley Terrane. Minor structures associated with the Highland Boundary Fault zone have been attributed to sinistral transpression. Both these major faults, together with the Southern Uplands Fault, are attributed to the late Silurian to early Devonian collision between Laurentia and Avalonia.

The Southern Uplands

This zone consists of several fault-bounded packages of steeply dipping Ordovician to Silurian strata bounded in the north by the Southern Uplands Fault, which has experienced at least 10km of sinistral strike-slip displacement. The beds in each individual package become younger northwards, although the more southeasterly packages contain younger material. The individual successions in the north contain at their base early Ordovician basalts and cherts overlain by black shales, succeeded by sedimentary sequences dominated by greywacke turbidites. The steeply inclined strata trend uniformly NE–SW and are affected by asymmetric upright folds. The shales possess a slaty cleavage.

The set of faults that occur throughout the belt are considered to represent steepened thrusts that have been re-activated in a strike-slip sense. The total sinistral offset across the zone is probably considerable and took place during the closing stages of the Caledonian orogeny along with the other strike-slip faults further north.

The Southern Uplands zone is bounded on its southeastern side by the Iapetus suture, which is not exposed, but has been seismically imaged as a major discontinuity inclined at a moderate angle north-westwards, and lies at a depth of around 12km in the central part of the zone. The basement beneath the suture is interpreted as part of the Avalonian Terrane. This zone is represented in Ireland by the Longford Down massif. The Southern Uplands has long been regarded as an accretionary prism, formed above a NW-dipping subduction zone at the northwestern margin of Avalonia (Fig. 14.7.5).

The Lake District zone

The Caledonian rocks of the English Lake District consist of a sequence of Ordovician arc-type volcanics, succeeded by Silurian marine deposits; these rest on a late Precambrian basement, which is exposed in Anglesey and NW Wales. The southwestern continuation of this zone in Ireland is represented by the Leinster massif. The zone is interpreted as an Ordovician volcanic arc situated at the northern margin of Avalonia. These Lower Palaeozoic rocks were deformed and subjected to slate-grade metamorphism in the early Devonian, at the same time as those of the Welsh Basin, and reflect the final collision of Laurentia and Avalonia.

The Welsh Basin

The Lower Palaeozoic rocks of Wales have been intensively studied by generations of geologists and were regarded as an example of a 'eugeosyncline' in the 1930s. They comprise around 10km of Cambrian to Silurian sediments, including a large proportion of turbidites. Volcanics are an important constituent, especially in the northwest, in Snowdonia. The rocks of the zone were deformed in the early Devonian; tight folds with associated slaty cleavage in the north give way to more gentle folds in the southeast. The zone has been interpreted as a back-arc basin situated behind the Lake District arc on thinned Avalonian crust (Fig. 14.7.4–5).

Tectonic history

This is summarised in Figure 14.7. Note that profiles 1–3 relate to the northern part of the British Isles (i.e. the Laurentia–Baltica relationship) and profiles 4 and 5 to the southern (the Laurentia–Avalonia relationship).

1 **Late Proterozoic–mid-Ordovician**. Deposition of shelf and continental slope sedimentary sequences on the thinned margin of the Laurentian continent.
2 **Mid-Ordovician**. The Grampian event: NW-directed overfolding and thrusting attributed to the

emplacement of a volcanic arc terrane over the Grampian Highlands. A similar event occurred in Scandinavia, involving eastwards overthrusting onto the margin of Baltica.

3 **Mid–late Silurian**. The Scandian event: refolding of Grampian and Northern Highlands, and thrusting of the Moine Thrust Belt, attributed to collision with the Midland Valley volcanic arc terrane and the convergence and collision with Baltica.

4 **Late Silurian–early Devonian**. Folding and thrusting in the Southern Uplands, Lake District and Welsh Basin, followed by sinistral strike-slip faulting, attributed to the collision between Laurentia and Avalonia.

Upper Palaeozoic Belts

The three main collisional belts that formed during the Upper Palaeozoic are shown in brown in Figure 14.1: the Appalachian–Ouachita Belt along the eastern and southern margins of North America; the Urals Belt along the eastern side of Baltica; and the Central Asian Belt between Siberia and Kazakhstania. The Northern Appalachians, from Newfoundland to New York, are part of the Caledonides (see Fig. 14.6) and experienced little deformation after the Acadian orogenic event – the Appalachian equivalent of the main Caledonian (late Silurian to early Devonian) orogenic phase. The Ouachita Belt and the Southern Appalachians, together with its counterpart in Northwest Africa, form the western end of a broad Upper Palaeozoic orogenic belt that extends across southern Europe to the western shores of the Black Sea. The European sector of this belt experienced further tectonic activity in the Alpine Orogeny with the collision between the Eurasian and African–Arabian plates (see chapters 4–6).

The Southern Appalachians and the Ouachita Belt

The Southern Appalachian mountain chain extends in a southwesterly direction from the Hudson River in the north to central Alabama in the south, passing through the States of Pennsylvania, Virginia, West Virginia, North Carolina and Georgia – a distance of about 2000km (Fig. 14.9). The mountain belt is narrower in the north but broadens to around 200km wide in the south, where some of the higher peaks are. The highest summit is Mount Mitchell (2037m), in the Black Mountains of North Carolina – part of the Blue Ridge Province. The mountain belt consists of numerous parallel ranges and

individual ridges, many of which reflect the geological structure in a direct way, such that individual folds can be easily traced in aerial views.

Tectonic overview

While the Caledonian Orogeny was affecting the Northern Appalachians, the British Isles and Norway, sedimentary deposition continued on the thinned passive margin of the Laurentian continent further south, which was not seriously disturbed until the Alleghenian Orogeny during the Permian, which culminated in the collision between Laurentia and Africa, resulting in the formation of the Southern Appalachian Belt.

The Southern Appalachian Belt consists of four main tectonic zones: the Appalachian Basin, the Valley and Ridge Province, the Blue Ridge Province and the Piedmont Zone (Fig. 14.9).

The Appalachian Basin

This zone is a foreland basin situated on the passive margin of the Laurentian continent, containing a sedimentary succession consisting mainly of Carboniferous strata, which are undeformed in the main part of the basin but gently folded near the eastern margin. These Carboniferous rocks are an important source of oil and gas deposits.

The Valley and Ridge Province

This zone contains strata ranging from Cambrian to early Permian without any appreciable break (i.e. neither the Taconic nor the Acadian orogenies of the Northern Appalachians has affected them). This Palaeozoic sequence has experienced folding and thrusting during the Permian in the Alleghenian Orogeny – the North American equivalent of the Hercynian Orogeny of Europe. The zone has long been used as a type example of a thin-skinned fold-thrust belt.

The Blue Ridge Province

This zone consists of a mid-Proterozoic (Grenville) crystalline basement with a late Proterozoic to early Palaeozoic sedimentary cover. The zone has been thrust westwards over the Valley and Ridge Province for a distance of over 240km along a major low-angle fault.

The Piedmont Province

This zone occupies the eastern foothills of the Appalachian Mountains and the Coastal Plain, where it is obscured by younger cover. It consists of a deformed and

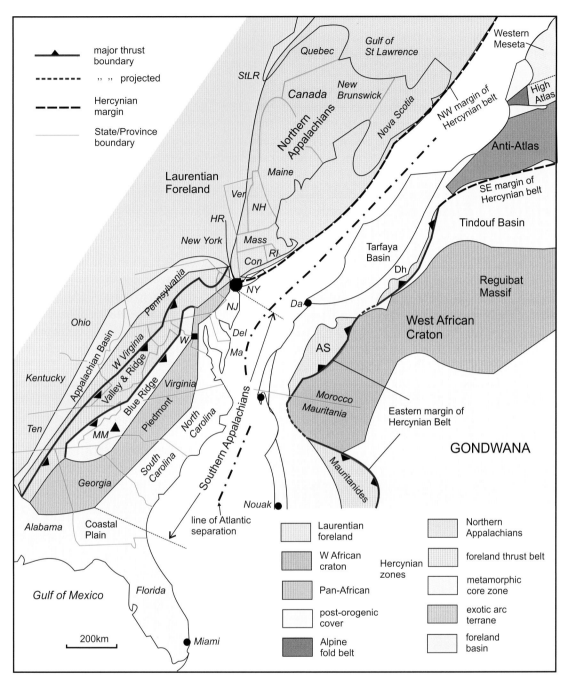

Figure 14.9 The Southern Appalachians and the Hercynian belts of NW Africa. Simplified map of the Hercynian tectonic belts on the opposite sides of the Atlantic Ocean, restored to their approximate pre-opening position. AS, Adrar Souttouf Belt; Con, Connecticut; Da, Dakhla; Del, Delaware; Dh, Dhlou; HR, Hudson River; Ma, Maryland; Mas, Massachussets; MM, Mt Mitchell; NH, New Hampshire; NJ, New Jersey; Nouak, Nouakchott; NY, New York; RI, Rhode Island; St LR, St Lawrence River; Ten, Tennessee; Ver, Vermont; W, Washington DC. After Michard *et al.*, 2008 (NW Africa); and US Geological Survey (Appalachians).

metamorphosed pre-Carboniferous complex, intruded by granitic and gabbroic plutons of mid-Carboniferous to early Permian age, and is interpreted as a volcanic arc. The zone includes Lower Palaeozoic greywackes and slates containing a European-type fauna comparable

with the Welsh slate belt. The Piedmont Province is interpreted as a group of exotic terranes of probable Gondwanan origin, similar to the Avalonian Terrane of the northern Appalachians, accreted to the Laurentian continent during the Alleghenian orogeny.

West Africa

The effects of the Hercynian Orogeny are preserved in several places along the western and north-western margins of Africa, as a result of the Gondwana–Laurasia collision, which has resulted in pieces of the eastern rim of the Appalachian orogen having been emplaced onto the margin of the West African Craton. In Western Mauretania and southwestern Morocco, reworked Precambrian basement has been thrust eastwards over the Reguibat Shield, part of the Archaean West African Craton, but Alleghenian deformation is relatively weak. There are three separate belts, from south to north: the Mauretanides in Western Mauretania, and the Adrar Souttouf and Dhlou belts in Southwest Morocco.

Further north, in Northwest Morocco, the Alleghenian orogeny has affected both the Western Meseta and Anti-Atlas regions. In the Western Meseta, the Precambrian basement, already strongly deformed and metamorphosed during the late Precambrian to early Cambrian Pan-African event, has been affected by further deformation and metamorphism of late Carboniferous age and intruded by granitic plutons. This belt, which is the Hercynian counterpart of the Northern Appalachians, is cut off on its southern side by the Alpine Atlas Belt. The Pan-African basement of the Anti-Atlas Belt on the southern side of the Alpine Atlas is also affected by Alleghenian deformation, but to a lesser extent, and is bounded to the south by post-Palaeozoic cover. Figure 14.9 shows that the Hercynian Orogenic Belt extended north-eastwards from the northern end of the Southern Appalachians and crossed to North Africa to include the Atlas region, but left the Northern Appalachians relatively unscathed on its northern side.

The Ouachita Belt

The Ouachita Belt (Fig. 14.10) forms a northwardly convex arc following the Ouachita Mountains of Arkansas and Oklahoma, and a westward continuation of the belt forms the Llano, Marathon and Solitaro Uplifts of Texas. The topography is relatively subdued – the highest mountain in the Ouachitas is only 839m high.

In geological terms, the Ouachita–Marathon Belt is a thin-skinned foreland fold-thrust belt developed on the southern passive margin of the Laurentian continent. The fold belt includes a typical passive-margin sedimentary sequence of Ordovician to Lower Carboniferous age. In the Upper Carboniferous (Mississippian), thick flysch deposits heralded the approach of a continental terrane from the south fringed by a south-directed subduction zone. This terrane is completely obscured by the post-orogenic cover of the coastal plain, but is considered to have finally collided with Laurentia in the late Upper Carboniferous to early Permian.

The Variscan Orogenic Belt

The Hercynian Orogenic Belt extends eastwards from the Appalachians and NW Africa, as shown in Figure 14.9, to form a wide zone in central and southern Europe,

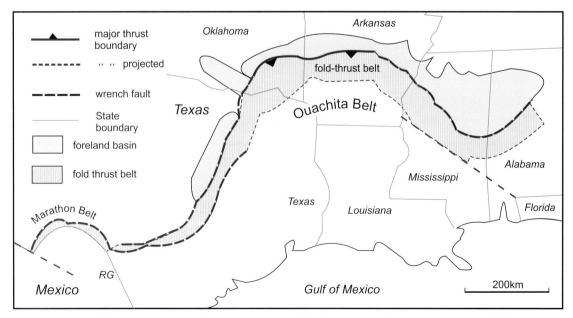

Figure 14.10 The Ouachita–Marathon Fold Belt. RG, Rio Grande. After Hickman *et al.*, 2009.

where it is known as the Variscan (see Fig. 14.1). The Variscan Belt begins at the western side of the Iberian Peninsula, and ends at the western margin of Baltica in Poland – a distance of around 2500km. It is cut off on its southern side by the younger Alpine Belt. Exposures of Variscan rocks in Europe are limited to upstanding massifs, such as the Massif Central in France and the Schwarzvald in Germany, showing generally low relief – typically less than 1500m – and surrounded by Mesozoic and Cenozoic sedimentary deposits.

Tectonic overview

The Variscan Orogenic Belt is a complex assemblage of terranes that have accreted to the southern margin of Laurussia (the combined continent of Laurentia plus Baltica) during the Devonian and Carboniferous (see Fig. 14.1). In West and Central Europe, the northern boundary of the orogenic belt, the Variscan Front, is the frontal thrust of a foreland thrust belt known as the Rheno-Hercynian Zone (RHZ), which can be traced from southern Ireland and southwest England across the Ardennes to the Rhenisches Schiefegebirge and the Harz Mountains in Germany (Fig. 14.11). The RHZ has been thrust across the southern passive margin of the Avalonian foreland, which extends beneath it at least to the southern margin of the RHZ and possibly beyond.

To the south of the RHZ is a series of exotic terranes collectively termed the Armorican Terrane Assemblage (ATA). The various units of the ATA are exposed in several Palaeozoic massifs that are now separated by Mesozoic and Cenozoic cover. Although the massifs have been extensively studied, the link between them is somewhat speculative. Figure 14.11 shows the simplest correlation between the terranes, but there are other possible interpretations. Several Gondwana-derived continental terranes have been recognised, of which the best preserved are Armorica and Perunica (otherwise known as Bohemia). Other pieces of Gondwanan

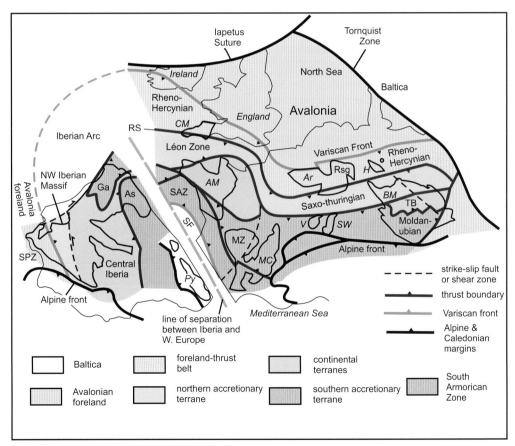

Figure 14.11 The Variscan Belt of Europe. Variscan Massifs: AM, Armorican; Ar, Ardennes; BM, Bohemian; CM, Cornubian; H, Harz Mountains; MC, Massif Central; Rsg, Rhenisches Schiefergebirge; SW, Schwarzwald; V, Vosges. Variscan Terranes: As, Asturias; Ga, Galician; MZ, Mauges; SAZ, South Armorican; SF, Southern Foreland (Gondwana); SPZ, South Portuguese; TB, Tepla-Barrandian. Py, Pyrenees; RS, Rheic Suture. Iberia is rotated to its approximate pre-Mesozoic position. After Ballèvre *et al.*, 2009; and Martinez-Catalan *et al.*, 2009.

continental crust, together with several volcanic arc terranes, have been incorporated within the complex accretionary regions to the north and south of the two major continental blocks.

Armorica and Perunica

The name 'Armorica' was traditionally attached to the Armorican Massif of Northwest France, but its use has been expanded to mean the large continental terrane underlying much of Western Europe, including the Iberian Peninsula and most of France. This terrane consists of a late Proterozoic to early Cambrian 'Pan-African' basement (known in Europe as the 'Cadomian') with a Lower Palaeozoic sedimentary cover originating in northern Gondwana. Its Cambrian to Devonian faunas are of Gondwanan affinities, and the terrane is believed to have been detached from Gondwana at the end of the Silurian when the Palaeo-Tethys Ocean opened (see Fig. 14.3B).

The other main crustal addition to Europe at this time, Perunica, includes the central part of the Bohemian Massif in the present Czech Republic, and is known there as the Teplá-Barrandian Zone (Fig. 14.11). This zone also contains a Cadomian basement, and is considered to have split from Gondwana and travelled across the Rheic Ocean separately during the Devonian (see Fig. 14.3B). Late Silurian faunas in the massif, like those of Armorica, have Gondwanan affinities and differ from those of Northern Europe.

The accretionary suture zones

The two continental terranes, Perunica and Armorica, are bounded on their northern and southern sides by complex zones containing both continental and oceanic material, and which contain the sutures where the intervening ocean basins have been consumed. The suture zone to the north is known as the Léon Zone in the west and the Saxothuringian Zone in the east. The southern suture zone is known as the Mauges Zone in France and the Moldanubian Zone in the Bohemian Massif.

South and west of the Mauges Zone lies the South Armorican Zone, which has been interpreted as part of the Gondwanan foreland. It is here that the first contact with the main Gondwana continent probably occurred at the end of the Carboniferous (see Fig. 14.1). Each of these zones can be traced into the northwestern part of the Iberian Massif, where they curve through more than 90º in a structure known as the Iberian Arc.

A number of Variscan terranes also occur within the Alpine Orogenic Belt in southern Europe, and along the southern margin of Baltica north of the Alpine Front. These terranes include sections of the Laurussian continental foreland, some of which have been involved in the Alpine Orogeny, and independent terranes that have been added to the Laurussian plate during the Mesozoic.

Tectonic evolution of the Variscides

Several continental terranes including Armorica, Teplá-Barrandia and Central Iberia separated from Gondwana during the Lower Palaeozoic, this separation probably promoted by back-arc spreading from a circum-Gondwana subduction zone.

During the Silurian, a volcanic arc formed within the Rheic Ocean caused by a north-dipping subduction zone, which resulted in pulling the ATA terranes away from Gondwana towards Laurussia. In the early Devonian, the Rheno-Hercynian back-arc basin formed above this subduction zone and eventually spread onto the Avalonian passive margin.

Continued expansion through the Devonian of the Palaeo-Tethys Ocean between the ATA and Gondwana resulted in the closure of the Rheic ocean basin and the subsequent amalgamation of the northern terranes of the ATA. The convergence direction of the combined terranes was oblique to the Tornquist Zone at the margin of Baltica, resulting in a combination of clockwise rotation and dextral shear in the marginal parts of the ATA. Continued expansion of the Palaeo-Tethys Ocean into the early Carboniferous caused the development of a subduction zone at the southern margin of Armorica and Teplá-Barrandia, leading to the development of the Mauges–Moldanubian accretionary complex.

During the mid-Carboniferous there was a change in convergence direction between Gondwana and Laurussia causing Gondwana to rotate anti-clockwise such that North Africa now faced North America, and the final collision took place between the NW African part of Gondwana and the section of Laurussia from the Appalachians to Iberia (see Fig. 14.1). Tectonic activity in more easterly sectors of the Central European Variscides was revived during the Alpine Orogeny (see chapters 4–6).

The Urals and Central Asia

Two further orogenic belts of Hercynian age complete the Palaeozoic assemblage that stitched Pangaea together in the Permian: the Uralides and the Central Asian Belt.

These belts are the result of the accretion of volcanic arcs, oceanic terranes and micro-continents to the passive margins of Siberia in the north, Baltica in the west, and the Tarim and North China continental blocks in the south over a long period of time, commencing in the late Proterozoic and ending with final collision in the late Carboniferous through to the early Mesozoic.

During the early Palaeozoic, the Aegir and Turkestan Oceans separated Siberia both from Baltica and from a number of micro-continental blocks: Tarim, North China (Sino-Korea) and several other terranes, which collectively formed Kazakhstania (Fig. 14.3B). The (present) eastern border of Baltica and the southern and western borders of Siberia seem to have behaved as passive margins until the Carboniferous Period and accumulated platform sediments up until that point. However, within the Aegir and Turkestan Oceans,

volcanic island arcs developed, fringed by subduction zones, and these ultimately became caught up in the late Carboniferous to early Mesozoic collision between the larger continents.

Figure 14.12 is a simplified map showing the complex assemblage of terranes making up a roughly triangular area sandwiched between Baltica, Siberia, and the Tarim and North China continental blocks. The western part of this assemblage, forming the separate continent of Kazakhstania, was completed in the late Silurian, and consists of Lower Palaeozoic subduction–accretion complexes juxtaposed with elongate bands and lensoid masses of continental crust of Gondwanan derivation. The Tarim and North China Blocks, which joined the assemblage in the Permian and Triassic respectively, are likewise founded on Gondwanan Precambrian basement.

This assemblage is bounded in the east by a large

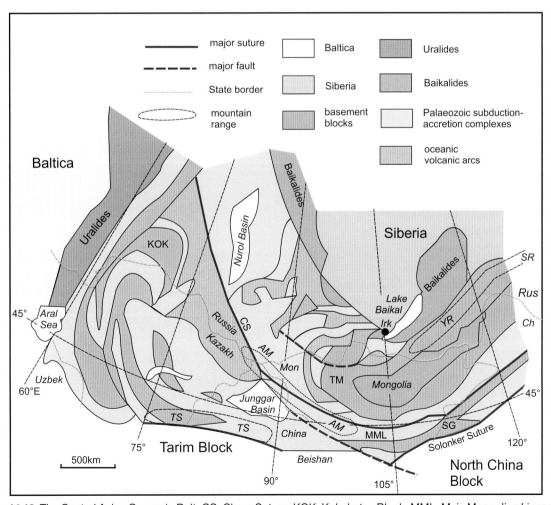

Figure 14.12 The Central Asian Orogenic Belt. CS, Chara Suture; KOK, Kokchetav Block; MML, Main Mongolian Lineament; SG, South Gobi Block; TM, Tuva-Mongol Massif. Geographic names: AM, Altai Mountains; Ch, China; Mon, Mongolia; SR, Stanovoy Range; Rus, Russia; TS, Tian Shan (Mts); YR, Yablonovy Range. After Windley *et al.*, 2007.

volcanic arc terrane that extends for over 1500km from the eastern margin of the Uralides to the Junggar Basin in Northwest China. The eastern side of this terrane is defined by the Chara Suture, which contains ophiolites and tectonic mélanges, and marks the eastern boundary of Kazakhstania, separating it from the terranes that have accreted to Siberia. The southern part of the Chara Suture and its continuation through Mongolia, known as the Main Mongolian Lineament, runs through the Altai Mountain Range. The southern margin of Kazakhstania lies on the southern side of the Tian Shan Range, and is defined by another suture separating it from the Tarim Block to the south.

The Uralides

The Ural Mountains form the eastern boundary of the European sub-continent and extend for a total distance of over 3,200km from the City of Orenburg near the Kazakhstan border to the northern end of the Novaya Zemlya islands in the Arctic Ocean. The mountain range is relatively narrow over most of its length but broadens to around 300km in width around Magnitogorsk, where the highest peaks are situated – the highest being Gora Yamantau at 1638m. Formed during the late Palaeozoic to early Mesozoic Uralian Orogeny, the Ural Range is the only component of the Hercynian orogenic system to offer a complete section across a collisional orogenic belt between two major continents, and in that sense is comparable with the more recent Himalayan belt.

Plate-tectonic context

The origins of the Uralian belt lie in a sequence of plate-tectonic movements that took place through the Silurian and Devonian Periods. Figure 14.3B shows a wide ocean, known as the Aegir Ocean, separating Baltica (then part of Laurussia) from the Siberian continent, with several small terranes between them. At that time, and throughout the early Carboniferous, the (present) eastern border of Baltica was a passive margin overlain by a wide shallow-marine shelf. By the late Silurian, several of the small terranes had combined to form the continent of Kazakhstania, which, in the mid-Carboniferous, collided with the eastern margin of Baltica, followed in the Permian Period by collision with Siberia to the north (see Fig. 14.1).

Main tectonic features of the Uralian Belt

The Uralian Orogenic Belt (Figure 14.13) follows the Ural Mountains that divide European Russia from Siberia. The orogenic belt divides into two branches in the north, separated by the Pechora Basin, the western branch following the late Proterozoic Timan Belt. Further south, the belt varies from about 200km to 600km in width, bounded in the west by the Uralian Foredeep Basin and in the east by the West Siberian Basin. At its southern end, the belt disappears beneath post-orogenic cover of the Peri-Caspian Basin.

The southern part of the Uralian belt can be divided into six separate tectonic zones, from west to east: the Uralian Foredeep, West Uralian, Central Uralian, Magnitogorsk, East Uralian and Trans-Uralian Zones.

The Uralian Foredeep Basin

This developed during the Carboniferous, and occupies the passive margin of Laurussia (i.e. originally Baltica). It grades westwards into a shallow-marine, mostly carbonate, shelf and is bounded in the east by the main boundary thrust of the Uralian Orogenic Belt. It contains continental clastic molasse deposits derived from the rising Ural Mountains. To the south, it is replaced by the shallow-marine deposits of the Peri-Caspian Basin, and to the north it divides into two branches, respectively following the western and eastern branches of the orogenic belt.

The West Uralian Zone

This is a typical foreland fold-thrust belt, and involves basement and sedimentary cover belonging to the East European Craton. The principal structure in the southern part of the orogen, illustrated in Figure 14.13B, is the large Bashkirian Anticlinorium, which exposes Precambrian Laurussian basement in its core. The thrusts in this zone dip uniformly eastwards, and end on a gently inclined mid-crustal detachment thrust which eventually meets the Main Uralian Fault (or Thrust) and thereby descends to the base of the crust at a depth of around 55km.

The Central Uralian Zone

This relatively narrow zone consists of folded and thrust high-grade metamorphic rocks (presumed to be part of the Laurussian basement) with a sedimentary cover of Carboniferous flysch deposits. The eastern part of the zone is occupied by a band of basic and ultrabasic ophiolites. The zone is bounded in the east by the Main Uralian Thrust, which forms the eastern margin of Laurussia, the presence of the ophiolites indicating that it marks a suture zone separating the Laurussian Plate on its western side from various exotic Asian terranes

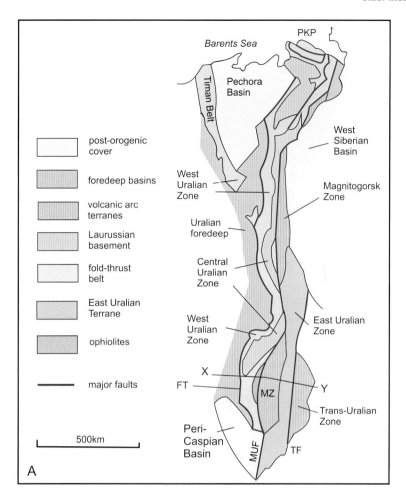

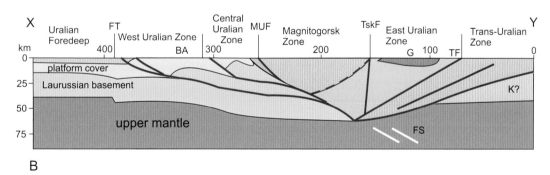

Figure 14.13 The Uralian Orogenic Belt. Simplified map (**A**) and cross-section (**B**) showing the main tectonic features. BA, Bashkirian anticlinorium; FS, fossil subduction zone?; FT, Frontal Thrust; G, granite intrusion; K?, Kazakhstania basement?; MUF, Main Uralian Fault; MZ, Magnitogorsk Zone; PKP, Pay-Khoi Peninsula; TF, Trans-Uralian Fault; TskF, Toisks Fault. A, after Juhlin *et al.*, 1998; B, after Berzin *et al.*, 1996.

on its eastern side. The thrust descends to the base of the crust at a depth of about 55km, and traces of its continuation through the upper mantle are believed to mark the former position of the east-dipping subduction zone that descended beneath the Magnitogorsk terrane to the east (Fig. 14.13B).

The Magnitogorsk Zone

This zone represents an oceanic volcanic arc complex, folded in the west and cut by west-dipping thrusts, in contrast to the east-dipping structures west of the suture. The zone is bounded on its eastern side by the East Uralian Zone; the nature of the boundary itself is uncertain but presumed to be a thrust. A major strike-slip fault, the Toistsk Fault, outcrops close to the boundary. The city of Magnitogorsk hosts a steelworks that was once the largest in Europe and was based on iron ore from Magnitnaya, the local 'magnetic mountain'.

The East Uralian Zone

This zone consists of metamorphic basement rocks belonging to a small continental terrane and is intruded by numerous granitic bodies of typical magmatic arc type. It is bounded in the east by a major west-dipping thrust, the Trans-Uralian Fault, which separates it from the Trans-Uralian Zone.

The Trans-Uralian Zone

This zone consists of a Palaeozoic volcanic arc complex. It is regarded as a separate arc terrane accreted to the East Uralian zone along a west-dipping subduction zone that is marked by sporadic ophiolite bodies. This zone is only recognised in the south of the Urals belt; its northerly continuation is obscured beneath the West Siberian Basin. The presence of a strong west-dipping seismic reflector beneath the Trans-Uralian Zone probably marks the top of an underthrust continental terrane (part of Kazakhstania).

Tectonic evolution

1 During the Silurian Period, the wide Aegir Ocean separated Baltica (Laurussia) from the nearest continental plate, Siberia. Within this ocean were several continental terranes and oceanic volcanic arcs, some of which, by the end of the Silurian, had amalgamated to form the continent of Kazakhstania. Between Kazakhstania and Laurussia were the Magnitogorsk volcanic arc and another continental terrane, which became the East Uralian Zone.

2 In the late Devonian, the Magnitogorsk Arc collided with the (present) eastern passive margin of Laurussia, and was thrust over it. A west-dipping subduction zone then formed beneath the Magnitogorsk Arc, bringing the East Uralian Terrane closer to the developing orogen. Another oceanic volcanic arc, which would become the Trans-Uralian Zone, formed above a second west-dipping subduction zone between the East Uralian Terrane and Kazakhstania.

3 In the early Carboniferous, the East Uralian Terrane underthrust the Magnitogorsk Zone, causing deformation and uplift of the orogenic belt to commence. Kazakhstania continued to approach Laurussia as the intervening oceanic plate was subducted beneath the Trans-Uralian Arc.

4 The mid- to late Carboniferous saw the climax of the Uralian Orogeny in the south; the Trans-Uralian Terrane collided with and underthrust the East Uralian Terrane, which in turn met and was underthrust by the leading edge of the Kazakhstanian assemblage, causing further compressive deformation throughout the orogeny and spreading into the Laurussian foreland to produce the foreland fold-thrust belt. The orogenic belt experienced considerable crustal thickening; even at the present day, the base of the crust beneath the central part of the orogen is still at more than 55km depth, compared to around 40km in the adjacent Laurussian platform.

5 Further north, commencing in the Permian and lasting into the early Jurassic, the Siberian continent collided with the northern sector of the Uralian belt, including Novaya Zemlya and the Taimyr Peninsula.

The Central Asian Belt

The Central Asian Orogenic Belt extends from the Ural Mountains in the west to the coast of the Sea of Okhotsk in eastern Siberia, a distance of around 6000km and nearly 80° of longitude. Much of the western part of the orogenic belt, in Central Kazakhstan, is relatively low-lying, but from Longitude 70°E a series of high mountain ranges stretches across eastern Kazakhstan, Mongolia and Northern China in a broad belt around 1000km across. These include the Tian Shan ranges, already discussed in chapter 8, together with the Altai Mountains in Western Mongolia, and the Yablonovyy and Stanovoy ranges along the China–Siberia border. The highest peak in the Altai Range is the 4506m-high Mt. Belukha; the eastern ranges are lower, below 3000m.

Tectonic overview

The terrane assemblage fringing Siberia forms a broad arc around the western side of the Siberian craton to Lake Baikal, then strikes north-eastwards (see Fig. 14.12). The Archaean–early Proterozoic core of Siberia is ringed by a late Proterozoic to early Cambrian orogenic belt known as the Baikalides. Around this margin are wrapped a complex array of terranes, similar to those composing Kazakhstania, consisting of continental blocks, oceanic volcanic arc terranes and subduction–accretion complexes ranging in age from latest Proterozoic to Carboniferous.

On the southeastern side of Siberia, an elongate block of Archaean to late Proterozoic basement, believed to be of Gondwanan origin, together with its platform cover, extends for over 2000km in a northeasterly direction. The northern side of this massif lies along the Yabonovyy and Stanovoy mountain ranges, and the massif is bordered on its south side by Palaeozoic subduction–accretion complexes and another basement terrane, the South Gobi Block, on the south side of which is the Solonker Suture, separating the accretionary belt from the North China Craton, which joined in the late Triassic.

Sinistral movements on several major strike-slip faults, including the Chara Suture and the Main Mongolian Lineament, which are curved around parallel to the margin of the Siberian Craton, are attributed to late Hercynian anti-clockwise rotation of Siberia relative to both Baltica and the southern cratons.

Orogeny in the Precambrian

It is generally believed that plate-tectonic processes essentially similar to those discussed above in relation to the Phanerozoic orogenic belts operated also in the Proterozoic, but there has been considerable debate over whether that is equally true for the Archaean. The consensus now is that, although there may be differences in detail – for example in the rate of heat flow through the crust, the thickness of the lithosphere, and the temperature and composition of magmas – the earliest crust known was produced by the same kinds of subduction–accretion processes that have been discussed for the Phanerozoic.

However, the major difference between investigations of the Phanerozoic belts and those formed during the Precambrian is that there is considerably more uncertainty regarding the sequence of plate-tectonic events responsible for the latter. In order to reconstruct these, it is necessary to restore the successive positions of the major continents before the Pangaea Supercontinent was formed, in the same way that was done in Figures 14.2 and 14.3 to end up with the reconstruction of Figure 14.1.

Most students of Precambrian tectonic history agree that a late Proterozoic supercontinent, known as Rodinia, broke up around the beginning of the Cambrian and that the various pieces rearranged themselves into the Pangaea Supercontinent. However, there is currently no generally agreed solution to the problem of how the various components of Rodinia fitted together. This problem is even greater for the postulated mid-Proterozoic supercontinent known as Nuna, and for the even more speculative end-Archaean supercontinent.

To conclude, therefore, although much is known about certain Precambrian orogenic belts such as the Grenville and Hudsonian belts of North America or the Sveco-Karelian and Sveco-Norwegian belts of Scandinavia, without more information concerning their relationships to their opposing continental hinterlands, it is not yet possible to reconstruct their tectonic history with any degree of certainty.

Glossary

It is assumed that the reader is familiar with the names of common rocks (e.g. basalt, granite, sandstone, etc.) and structures (e.g. fold, fault).

A

accretionary complex: an accumulation of **terranes** of varying nature (e.g. continental, oceanic, or volcanic-arc) added to a continental margin during the **subduction** process.

accretionary prism: an accumulation of sediments and volcanic debris occupying the ocean trench and continental slope at a **subduction** zone; these are often piled up in a series of folded and thrust slices (see Fig. 3.11).

active margin: (of a continent) characterised by a **subduction** zone.

allochthonous: of a tectonic unit (e.g. a **thrust** sheet), derived from elsewhere.

alkaline: magma or igneous rock characterised by relatively high proportions of alkalis compared with calcium.

andesite: fine-grained igneous rock, intermediate in composition between basalt and **rhyolite**; typical product of **subduction**-related magmas.

asthenosphere: the weak layer of the upper mantle, situated beneath the **lithosphere**, capable of solid-state flow, which enables the **plates** to move (see Figs 3.5, 3.8).

autochthonous: of a tectonic unit (e.g. a **thrust** sheet), derived locally.

B

back-arc basin: sedimentary basin situated on the upper plate of a subduction zone behind the volcanic arc (see Fig. 3.14).

batholith: a very large igneous body, typically composed predominantly of **calc-alkaline** rocks such as **granodiorite**, and consisting of many individual **plutons** intruded over a long period.

black smoker: active submarine volcanic vent: typically located on ocean ridges.

C

calc-alkaline: magma or igneous rock (e.g. basalt, **andesite** or granite) characterised by relatively high proportions of calcium compared with alkalis.

carbonate: sedimentary rock composed mainly of carbonates of calcium (limestone) and/or calcium-magnesium (dolostone).

channel flow: process of gravity-induced lateral flow of a layer of warm, ductile material within an **orogen**.

clastic: (of sediments) composed primarily of fragments of older rock.

continental drift: theory of the relative movements of the continents, popularised originally by Alfred Wegener and incorporated into the modern theory of **plate** tectonics.

core complex: see **metamorphic core complex**.

craton: stable part of a continent, not involved in contemporary **orogenic** activity.

crust: uppermost layer of the solid Earth, composed of a wide variety of rocks and varying in thickness from *c.*10km in the oceans to over 80km in **orogenic** belts; it rests on the much more uniform **mantle** (see Fig. 3.8).

D

décollement: process where a rock layer or tectonic unit becomes detached from its base and slides along a detachment surface.

dextral: (right-lateral): movement (e.g. along a fault) where the side opposite the observer moves to the right.

dyke: a sheet-like igneous body, typically with a vertical or steeply inclined attitude, generally discordant with the prevailing structure (e.g. bedding) of the host rock.

E

eugeosyncline: (obsolete): type of **geosyncline** characterised by deep-marine sediments and vulcanicity.

evaporite: sedimentary deposit formed by the evaporation of water containing soluble salts.

F

flysch: marine sediments (typically including **turbidites**) derived from an active mountain range or island arc.

fore-arc: region on the ocean-ward side of a volcanic arc (e.g. see Figs 9.3, 9.5).

foredeep basin: a sedimentary basin resulting from the depression of the continental **crust** due to the load of a rising **orogenic** belt; it contains a thick sequence of sediments derived from its erosion (e.g. see Fig. 4.9).

foreland: that part of the continental **crust** lying immediately adjacent to an **orogenic** belt, and which has not been significantly affected by it.

G

geosyncline: obsolete term used to describe a large elongate basin within the **crust** that gradually deepened and became filled with sediment.

Gondwana: (formerly Gondwanaland): a supercontinent that existed during Palaeozoic time, consisting of the continents of South America, Africa, India, Antarctica and Australia (see Figs 2.4, 2.5)

graben: fault-bounded depression, created by extension.

granodiorite: the **calc-alkaline** variety of granite; coarse-grained equivalent of **andesite**.

greywacke: type of poorly sorted **clastic** sediment composed of material of varying composition and size: typical of **turbidites**.

H

horst: uplifted fault-bounded block, created by compression.

I

imbricate: (structure, zone): where the same sequence of strata is repeated many times in successive **thrust** slices (e.g. see Fig. 14.8A).

isostasy: state of general gravitational equilibrium in which the extra weight of (e.g.) a mountain range is balanced by a deficiency of denser material beneath.

K

karst: type of terrain dominated by **carbonate** rocks characterised by distinctive landforms and drainage patterns caused by solution by rainwater.

klippe: outcrop of a **thrust** sheet that has been separated from the rest of the sheet by erosion (e.g. see Fig. 5.4B).

L

Laurasia: a supercontinent that existed during Upper Palaeozoic time, consisting of the greater parts of the continents of North America, Europe and Asia.

lherzolite: type of ultrabasic igneous rock capable of yielding basaltic magma on partial melting and thought to represent the composition of the upper **mantle**.

listric: (fault) where the fault angle becomes shallower at depth: typical of extensional structures.

lithosphere: the strong upper layer of the Earth, with an average thickness of about 100km, including the **crust** and part of the upper **mantle**; it consists of a number of **plates** that move over the weaker **asthenosphere** beneath.

M

mantle: that part of the Earth's interior between the **crust** and the core, composed mainly of rock with an ultrabasic composition (see Fig. 3.8).

mélange: rock unit, either sedimentary or tectonic, containing a mixture of material of varying provenance.

metamorphic core complex: interior part of an orogenic belt containing rocks that have been brought up from a considerable depth and have been intensely deformed and metamorphosed under high temperature and pressure.

miogeosyncline: (obsolete): type of **geosyncline** characterised by shallow-marine or continental shelf sediments and lacking volcanics.

molasse: non-marine sedimentary deposits (often red) derived from an active mountain range and deposited on continental crust.

N

nappe: a displaced tectonic unit resting on a **thrust**: a thrust sheet.

Neptunism: (obsolete) theory held by some eighteenth-century thinkers that all rocks (including crystalline igneous rocks such as granite) were precipitated from the great biblical flood.

O

obduction: process where a **terrane** or rock unit is thrust over an opposing plate at a **subduction** zone: i.e. the opposite of subduction.

ophiolite: rock unit or sequence interpreted as part of the oceanic **lithosphere**, consisting of basic and ultrabasic igneous material and oceanic sediments. A typical ophiolite should include oceanic lavas (e.g. **pillow basalts**), and both basic and ultrabasic intrusions (e.g. see Fig. 3.8A).

orogen, orogenic belt: part of the Earth's crust, typically linear, formed as a result of orogeny.

orogeny: the process of mountain building resulting from plate collision, involving crustal thickening, uplift and the formation of mountains.

P

palaeolatitude: the latitude of a rock or rock unit (e.g. a **terrane**) at a previous time period as determined by **palaeomagnetism**.

palaeolongitude: the longitude of a rock or rock unit (e.g. a **terrane**) at a previous time period (this cannot be determined by **palaeomagnetism**).

palaeomagnetism: the study of the magnetic properties of rocks: principally to determine the orientation of their inherited magnetic **palaeolatitude** and pole position.

Pangaea: the supercontinent, consisting of (almost) the whole continental landmass, which existed during much of Upper Palaeozoic time (see Fig. 14.1).

passive margin: (of a continent): lacking evidence of **subduction**.

pegmatite: very coarse-grained crystalline rock, typically granitic, of magmatic or metamorphic origin, often formed from volatile-rich fluids.

pericline: large-scale fold with an elongate oval outcrop pattern.

pillow basalt: basaltic lava flow exhibiting pillow-shaped or tube-like structures formed by rapid cooling in relatively shallow water.

plate: a relatively stable piece of the **lithosphere** that moves independently of adjoining plates; plate boundaries are where tectonic processes such as earthquakes and **orogeny** are concentrated.

plate tectonics: the theory that ascribes tectonic processes to the relative movement of the lithosphere **plates**.

plateau basalt: type of basalt forming voluminous flow sequences on the continents; typical of **plume**-related volcanism preceding continental separation.

plume (mantle): column of hot rising **mantle** material thought to be responsible for 'hotspots' in the **crust**.

pluton: large intrusive igneous body.

R

reverse fault: where the fault plane is inclined towards the upthrown side.

rhyolite: fine-grained igneous rock of granitic composition, found typically in the form of lava flows.

rift: extensional fault-bounded valley or trough.

S

sea-floor spreading: (obsolete) theory explaining **continental drift** by movements of the ocean **crust**.

seismogenic: (layer): part of the strong, brittle upper **crust** within which earthquakes originate.

shear zone: zone of ductile deformation, the deeper-level equivalent of a fault.

sill: a sheet-like igneous body, typically with a horizontal or gently inclined attitude, generally parallel to the prevailing structure (e.g. bedding) of the host rock.

sinistral: (left-lateral): of movement along a fault or **shear zone** etc., where the opposite side, as seen by an observer, moves to the left.

strike-slip: (fault, **shear zone** etc.) where the movement has taken place horizontally along the structure (see also **wrench fault** – Fig. 3.2).

subduction: the process whereby an oceanic **plate** descends into the **mantle** along a subduction zone; part of the subducted plate melts to give rise to a zone of volcanoes on the opposite (upper) plate (see Figs 3.5, 3.10, 3.11).

suture: line or surface along which separate crustal blocks, **terranes** or continents have been joined together as a result of the **subduction** of intervening oceanic material.

T

terrane: (micro-plate): piece of crust, smaller in scale than a **plate**, that has moved independently of adjoining crustal units and has experienced a different tectonic history from them.

thin-skinned: (deformation): where deformation, typically **thrust**-related, takes place within a relatively thin upper layer leaving the material beneath undisturbed.

tholeiite: type of **calc-alkaline** basalt characterised by relatively high proportions of silica; typical type of basalt forming the oceanic **crust**.

thrust: a (usually) gently inclined fault that has emplaced an older (or lower) rock unit above a younger (or upper) one (e.g. see Fig. 14.8A).

transform fault: a fault that forms part of a **plate** boundary where the plates on each side move in opposite directions, parallel to the trend of the fault.

transpression: combination of **strike-slip** and convergent (compressional) movement.

transtension: combination of **strike-slip** and divergent (extensional) movement.

trench roll-back: (also known as slab roll-back): process where the outcrop position of a **subduction** zone (i.e. the trench) moves backwards along the subducting **plate** over time, creating extensional conditions on the upper plate.

triple junction: where three **plate** boundaries meet.

turbidite: a deposit, formed by a **turbidity current,** and characterised by poorly sorted sediment of varying coarseness and composition.

turbidity current: a water current generated by gravity-induced flow, carrying large quantities of sediment in suspension.

W

window (tectonic): where erosion has created a gap in a **nappe** or **thrust** sheet exposing the rock sequence beneath (e.g. see Fig. 5.4).

wrench fault: a fault where the displacement is horizontal and parallel to the trend of the fault (see Fig. 3.2).

References and Further Reading

Chapter 2
References
Du Toit, A.L. (1937) *Our Wandering Continents. An Hypothesis of Continental Drifting.* London: Oliver & Boyd.

Geikie, Sir Archibald (1882) *Textbook of Geology.* London: Macmillan & Co.

Hess, H.H. (1962) History of ocean basins. In: A.E.J. Engel *et al.* (eds) *Petrologic studies: a volume in honour of A.F. Buddington.* Boulder, Colorado: Geological Society of America.

Holmes, A. (1913) *The Age of the Earth.* London: Harper & Brothers.

Holmes, A. (1929) Radioactivity and earth movements. *Transactions of the Geological Society of Glasgow* **18**, 559–606.

Holmes, A. (1944) *Principles of Physical Geology.* Edinburgh: Thomas Nelson & Sons.

Hutton, James (1788) Theory of the Earth; or an investigation of the laws observable in the composition, dissolution, and restoration of land upon the Globe. *Transactions of the Royal Society of Edinburgh* **1**, Part 2, pp. 209–304.

Hutton, James (1795) *Theory of the Earth.* Edinburgh: Creech.

Jeffreys, H. (1924) *The Earth, its origin, history and physical constitution.* Cambridge: Cambridge University Press.

Jeffreys, H. (1935) *Earthquakes and Mountains.* Cambridge: Cambridge University Press.

Lyell, Charles (1838–1865) *Elements of Geology.* London: John Murray.

Runcorn, S.K. (1962) Palaeomagnetic evidence for continental drift and its geophysical cause. In: S.K. Runcorn (ed.) *Continental Drift.* New York & London: Academic Press.

Suess, Eduard (1906) *Das Antlitz der Erde* (The Face of the Earth, translated H. Sollas) Oxford: Clarendon Press.

Taylor, F.B. (1910) Bearing of the Tertiary mountain belt on the origin of the Earth's plan. *Bulletin of the Geological Society of America* **21**, 179–226.

Umbgrove, J.H.F. (1950) *Symphony of the Earth.* The Hague: Martinus Nijhoff.

Wegener, A. (1922) *Die Enstehung der Kontinente und Ozeane* (The Origin of Continents and Oceans). Braunschweig: Vieweg.

Chapter 3
References
Le Pichon, X. (1968) Sea-floor spreading and continental drift. *Journal of Geophysical Research* **73**, 3661–3697.

McKenzie, D.P. and Parker, R.L. (1967) The North Pacific: an example of tectonics on a sphere. *Nature* **216**, 1276–1279.

Morgan, W.J. (1968) Rises, trenches, great faults and crustal blocks. *Journal of Geophysical Research* **73**, 1959–1982.

Further reading
Hallam, A. (1973) *A Revolution in the Earth Sciences: from Continental Drift to Plate Tectonics.* Oxford: Clarendon Press.

Chapter 4
Further reading
Alonso-Chaves, F.M., Soto, J.I., Orozco, M., Kilias, A.A. and Tranos, M.D. (2004) Tectonic evolution of the Betic Cordillera: an overview. *Bulletin of the Geological Society of Greece* **36**, 1598–1607.

Gibbons, W and Moreno, M.T. (2002) *The Geology of Spain.* The Geological Society, London.

Handy, M.R., Schmid, S.M., Bousquet, R., Kissling, E. and Bernoulli, D. (2010) Reconciling plate-tectonic reconstructions of Alpine Tethys with the geological–geophysical record of spreading and subduction in the Alps. *Earth Science Reviews* **102**, 121–158.

Chapter 5
References
Agassiz, L. (1840) *Etude sûr les glaciers.* Neuchatel: Jent et Gassmann.

Argand, E. (1916) Sur l'arc des Alpes occidentales. *Eclogae Geologicae Helvetiae* **14**, 145–191.

Collet, L.W. (1927) *The Structure of the Alps.* London: E. Arnold.

Heim, A. (1921) *Geologie der Schweiz.* Leipzig: Tauchnitz.

Staub, R. (1928) *Der Bewegungsmechanismus der Erde.* Berlin: Borntraeger.

Further reading
Patacca, E. and Scandone, P. (2007) Geology of the Southern Apennines. *Bollettino del Societa Geologia Italiana*, Special Issue **7**, 75–119.

Pfiffner, A. (2014) *Geology of the Alps.* Chichester: Wiley Blackwell.

Schmid, S.M., Fügenschuh, B., Kissling, E. and Schuster, R. (2004) Tectonic map and overall architecture of the Alpine Orogeny. *Eclogae geologicae Helvetica* **97**, 93–117.

Zeck, H.P. (1999) Alpine plate kinematics in the western

Mediterranean: a westwards-directed subduction regime followed by slab roll-back and slab detachment. In: B. Durand, L. Jolivet, F. Horvath and M. Séranne (eds) *The Mediterranean basins: Tertiary extension within the Alpine Orogen.* Geological Society of London, Special Publications **156**, 109–120.

Chapter 6
Further reading

Adamia, S., Zakariadze, G., Chkhotua, T., Sadradze, N., Tsereteli, N., Chabukiani, A. and Gventsadze, A. (2011) Geology of the Caucasus: a review. *Turkish Journal of Earth Sciences* **20**, 489–544.

Márton, E., Tischler, M., Csontos, L., Fügenschuh, B. and Schmid, S.M. (2007) The contact zone between the ALCAPA and Tisza–Dacia mega-tectonic units of Northern Romania in the light of new paleomagnetic data. *Swiss Journal of Geosciences* **100**, 1–16.

Okay, A.I. (2000) Geology of Turkey: a synopsis. *Anschnitt* **21**, 19–42.

Robertson, A.H.F. and Mountrakis, D. (eds) (2006) *Tectonic development of the Eastern Mediterranean Region.* Geological Society, London, Special Publications, 260.

Tari, V. (2002) Evolution of the northern and western Dinarides: a tectonostratigraphic approach. *EGU Stephan Mueller Special Publication* Series **1**, 223–236.

Chapter 7
Further reading

Leturmy, P. and Robin, C. (eds) (2010) *Tectonic and stratigraphic evolution of the Zagros and Makran during the Mesozoic–Cenozoic.* Geological Society, London, Special Publications, **330**.

Molinaro, M., Leturmy, P., Guezu, J.C., Frizon De Lamotte, D. and Eshraghi, S.A. (2005) The structure and kinematics of the south-eastern Zagros fold-thrust belt, Iran: from thin-skinned to thick-skinned tectonics. *Tectonics* **24**, TC3007.

Chapter 8
References

Molnar, P. and Tapponnier, P. (1975) Cenozoic tectonics of Asia: effects of a continental collision. *Science* **189** (4201), 419–426.

Further reading

Harrison, T.M. (2006) Did the Himalayan crystallines extrude partially molten from beneath the Tibetan Plateau? In: R.D. Law, M.P. Searle and L. Godin (eds) *Channel flow, ductile extrusion and exhumation in continental collision zones.* Geological Society of London, Special Publications **268**, 355–378.

Searle, M.P., Elliott, J.R., Phillips, R.J. *et al.* (2011) Crustal-lithospheric structure and continental extrusion of Tibet. *Journal of the Geological Society, London* **168**, 633–672.

Chapter 9
Further reading

Hall, R., Cottam, M.A. and Wilson, M.E.J. (eds) (2011) *The SE Asian gateway: history and tectonics of the Australia–Asia collision.* Geological Society, London, Special Publications, **355**.

Chapter 10
References

Miyashiro, A. (1961) Evolution of metamorphic belts. *Journal of Petrology* **2**, 277–311.

Further reading

Lallemand, S., Dominguez, S., Deschamps, A. and Liu, C-S. (2002) *Arc–continent collision in Taiwan: new marine observations and tectonic evolution.* Geological Society of America, Special Paper **358**, 189–213.

Larter, R.D. and Leat, P.T. (eds) (2003) *Intra-oceanic subduction systems: tectonic and magmatic processes.* Geological Society, London, Special Publications **219**.

Moreno, T., Wallis, S., Kojima, T and Gibbons, W. (2016) *The Geology of Japan.* Geological Society, London.

Chapter 11
References

Coney, P.J., Jones, D.L. and Monger, J.W.H. (1980) Cordilleran suspect terranes. *Nature* **288**, 329–333.

Further reading

Fitz-Diaz, E., Hudleston, P. and Tolson, G. (2011) Comparison of tectonic styles in the Mexican and Canadian Rocky Mountain Fold-thrust Belt. In: J. Poblet and R.J. Lisle (eds) *Kinematic evolution and structural styles of fold-thrust belts.* Geological Society, London, Special Publications, **349**, 149–167.

Moores, E.M. and Twiss, R.J. (1995) The Andes. In: *Tectonics.* New York: Freeman & Co.

Chapter 12
Further reading

James, K.H. (2013) Caribbean geology: extended and subsided continental crust sharing history with eastern North America, the Gulf of Mexico, the Yucatán Basin and northern South America. *Geoscience Canada* **40**, 1.

Moores, E.M. and Twiss, R.J. (1995) The North American Cordillera. In: *Tectonics.* New York: Freeman.

Moreno, T, and Gibbons, W. (2007) *The Geology of Chile.* The Geological Society, London.

Pfiffner, O.A. and Gonzales, L. (2013) Mesozoic–Cenozoic evolution of the western margin of South America: case study of the Peruvian Andes. *Geosciences* 2013, **3**, 262–310.

Chapter 13
References

Heezen, B.C. and Tharp, M. (1977) *World Ocean Floor Panorama*. NY: Marie Thorp Maps, 10976.

Wilson, J.T. (1963). A possible origin of the Hawaiian Islands. *Canadian Journal of Physics.* **41** (6), 863–870.

Further reading

Foulger, G.R. and Anderson, D.L. (2005) A cool model for the Iceland hotspot. *Journal of Volcanology and Geothermal research* **141**, 1–22.

Searle, R. (2015) *Mid-ocean Ridges*. Cambridge: Cambridge University Press.

Chapter 14
References

Peach, B.N., Horne, J., Gunn, W., Clough, C.T., Hinxman, L.W. and Teall, J.J.H. (1907) The geological structure of the northwest Highlands of Scotland. *Memoirs of the Geological Survey of Great Britain*.

Further reading

Ballèvre, M., Bosse, V., Ducassou, C. and Pitra, P. (2009) Palaeozoic history of the Armorican Massif: models for the tectonic evolution of the suture zones. *Comptes Rendus Geoscience* **341**, 174–201.

Cocks, L.R.M. and Torsvik, T.H. (2006) European geography in a global context from the Vendian to the end of the Palaeozoic. In: D.G. Gee and R. Stephenson (eds) *European lithosphere dynamics*. Geological Society of London, Memoirs **32**, 83–95.

Gee, D., Juhlin, C., Pascal, C. and Robinson, P. (2010) Collisional orogeny in the Scandinavian Caledonides. *Geologiska Föreningen i Stockholm Förhandlingar* **132**, 29–44.

Juhlin C., Friberg M., Echtler H., Hismatulin T., Rybalka A., Green A.G. and Ansorge J. (1998) Crustal structure of the Middle Urals: results from the ESRU experiments. *Tectonics* **17**(5), 710–725.

Leslie, A.G., Smith, M. and Soper, N.J. (2008) Laurentian margin evolution and the Caledonian Orogeny: a template for Scotland and East Greenland. In: A.K. Higgins, J.A. Gilotti and M.P. Smith (eds.) *The Greenland Caledonides: evolution of the northwest margin of Laurentia.* Geological Society of America, Memoirs **202**, 307–343.

Michard, A. *et al.* (2008) *Continental Evolution: the Geology of Morocco.* Lecture Notes in Earth Sciences **116**, Berlin Heidelberg: Springer-Verlag.

Roberts, D. (2003) The Scandinavian Caledonides: event chronology, palaeogeographic settings and likely modern analogues. *Tectonophysics* **365**, 283–299.

Trewin, N.H. (ed.) (2002) *The Geology of Scotland* (4th edition). Geological Society of London.

Windley, B.F., Alexeiev, D., Xiao, W., Kröner, A. and Badarch, G. (2007) Tectonic models for accretion of the Central Asian Orogenic Belt. *Journal of the Geological Society, London* **164**, 31–47.

Index

Page numbers in italic denote figures